BACK TO THE FUTURE
WITH RETRO GAMES

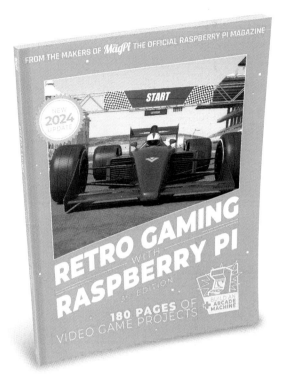

The 1980s and 1990s were a glorious era for gaming! In just twelve short years (1982-1994) we had the Sinclair Spectrum, Commodore 64, Amiga, and Atari ST; NES, SNES, Sega Master System, Sega Genesis/Mega Drive, and Saturn right up to the Sony PlayStation.

The pace of change from bitmapped graphics, through to sprite scaling and eventually 3D polygon graphics was breathtaking. We're still nursing sore thumbs from endless button-bashing.

Raspberry Pi makes low-cost, single-board computers that help kids (of all ages) to make and build with electronics. And one of the best things you can make is your own arcade machine or retro console. The new Raspberry Pi 5 computer is capable of emulating consoles all the way up to the Sega Dreamcast [the best console – Ed].

This book shows you, step-by-step, how to turn Raspberry Pi into several classic consoles and computers. Discover where to get brand new games from, and even how to start coding games. If you're brave: we'll show you how to build a full-sized arcade machine.

Rediscovering retro games is a fantastic hobby. You get all the thrill of nostalgia, and replay classic games that still hold up today, and you learn how computers and consoles work in the process.

Enjoy your trip back in time!

Lucy Hattersley-Haworth

FIND US ONLINE

EDITORIAL
Editor: **Lucy Hattersley-Haworth**
Features Editor: **Rob Zwetsloot**
Contributors: **David Crookes, PJ Evans, Rosie Hattersley, Nicola King, Phil King, KG Orphanides, Mark Vanstone**

DISTRIBUTION
Seymour Distribution Ltd
2 East Poultry Ave, London,
EC1A 9PT | **+44 (0)207 429 4000**

GET IN TOUCH

DESIGN
Critical Media: **criticalmedia.co.uk**
Head of Design: **Lee Allen**
Designers: **Sam Ribbits**
Illustrator: **Sam Alder, Dan Malone**

MAGAZINE SUBSCRIPTIONS
Unit 6, The Enterprise Centre,
Kelvin Lane, Manor Royal,
Crawley, West Sussex,
RH10 9PE | **+44 (0)207 429 4000**
magpi.cc/subscribe
magpi@subscriptionhelpline.co.uk

PUBLISHING
Publishing Director: **Brian Jepson**
brian.jepson@raspberrypi.com

Advertising: **Charlotte Milligan**
charlotte.milligan@raspberrypi.com
Tel: +44 (0)7725 368887

Director of Communications: **Liz Upton**
CEO: **Eben Upton**

CONTENTS

22

102 MAKE YOUR OWN GAMES

130 BUILD YOUR OWN RETRO MACHINES

SET UP YOUR SYSTEM

EVERYTHING YOU NEED TO GET UP AND RUNNING

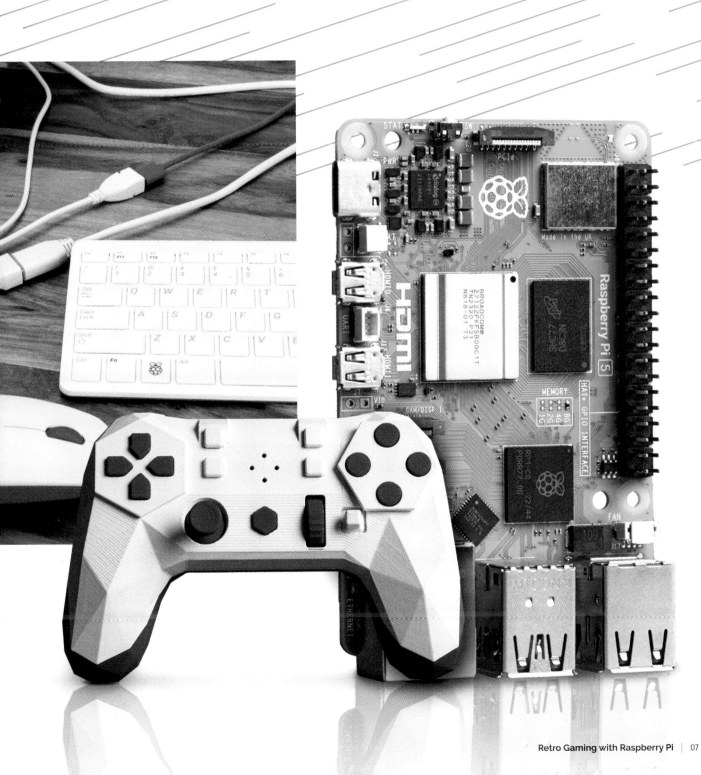

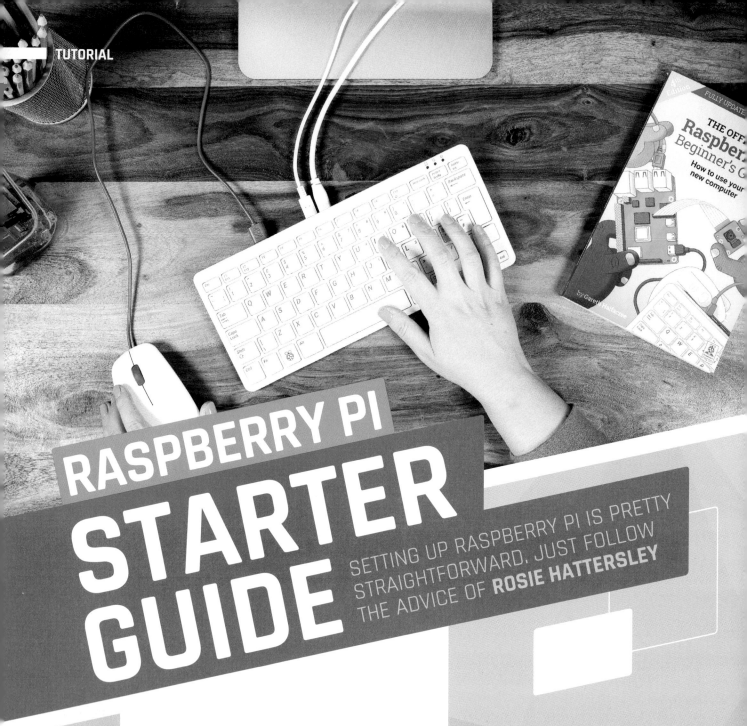

RASPBERRY PI
STARTER GUIDE

SETTING UP RASPBERRY PI IS PRETTY STRAIGHTFORWARD. JUST FOLLOW THE ADVICE OF **ROSIE HATTERSLEY**

Congratulations on becoming a Raspberry Pi explorer. We're sure you'll enjoy discovering a whole new world of computing and the chance to handcraft your own games, control your own robots and machines, and share your experiences with other Raspberry Pi fans.

Getting started won't take long: just corral the extra bits and bobs you need on our checklist. Useful additions include some headphones or speakers if you're keen on using Raspberry Pi as a media centre, or a gamepad for use as a retro games console.

To get set up, simply use your pre-written microSD card (or use Raspberry Pi Imager to set up a card) and connect all the cables. This guide will lead you through each step. You'll find the Raspberry Pi OS, including coding programs and office software, all available to use. After that, the world of digital making with Raspberry Pi awaits you.

What you need

All the bits and bobs you need to set up a Raspberry Pi computer

A Raspberry Pi

Whether you choose the new Raspberry Pi 5; or a Raspberry Pi 4, 400, 3B+, 3B; Raspberry Pi Zero or Zero 2 (or an older model of Raspberry Pi), basic setup is the same. All Raspberry Pi computers run from a microSD card, use a USB power supply, and feature the same operating systems, programs, and games.

8GB microSD card

You'll need a microSD card with a capacity of 8GB or greater. Your Raspberry Pi uses it to store games, programs, and boot the operating system. Many Raspberry Pi computer kits come with a card pre-written with Raspberry Pi OS. If you want to reuse an old card, you'll need a card reader: either USB or a microSD to full-sized SD (pictured).

Windows/Linux PC or Mac computer

You'll need a computer to write Raspberry Pi OS to the microSD card. It doesn't matter what operating system this computer runs, because it's just for installing the OS using the Raspberry Pi Imager app.

USB keyboard

Like any computer, you need a means to enter web addresses, type commands, and otherwise control Raspberry Pi. The Raspberry Pi 400 comes with its own keyboard. Raspberry Pi sells an official Keyboard and Hub (**magpi.cc/keyboard**) for other models.

USB mouse

A tethered mouse that physically attaches to your Raspberry Pi via a USB port is simplest and, unlike a Bluetooth version, is less likely to get lost just when you need it. Like the keyboard, we think it's best to perform the setup with a wired mouse. Raspberry Pi sells an Official Mouse (**magpi.cc/mouse**).

Power supply

Raspberry Pi 4, 400, and 5 need a USB Type-C power supply. Raspberry Pi sells power supplies (**magpi.cc/usbcpower**), which provide a reliable source of power – espeically for the power hungry Raspberry Pi 5. Raspberry Pi 1, 2, 3, and Zero models need a micro USB power supply (**magpi.cc/universalpower**).

Display and HDMI cable

A standard PC monitor is ideal, as the screen will be large enough to read comfortably. It needs to have an HDMI connection, as that's what's fitted on your Raspberry Pi board. Raspberry Pi 4, 400 and 5 can power two HDMI displays, but require a micro-HDMI to HDMI cable. Raspberry Pi 3B+ and 3A+ both use regular HDMI cables; Raspberry Pi Zero W needs a mini-HDMI to HDMI cable (or adapter).

USB hub

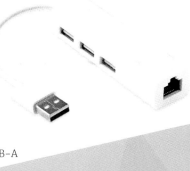

Raspberry Pi Zero and Model A boards have a single USB socket. To attach a keyboard and mouse (and other items), you should get a four-port USB hub (or use the official USB Keyboard and Hub which includes three ports). Instead of standard-size USB ports, Raspberry Pi Zero has a micro USB port (and usually comes with a micro USB to USB-A adapter).

SET UP RASPBERRY PI

Raspberry Pi 5 / 4 / 3B+ / 3 has plenty of connections, making it easy to set up

01 Hook up the keyboard

Connect a regular wired PC (or Mac) keyboard to one of the four larger USB-A sockets on a Raspberry Pi 5 / 4 / 3/3B+. It doesn't matter which USB-A socket you connect it to. It is possible to connect a Bluetooth keyboard, but it's much better to use a wired keyboard to start with.

02 Connect a mouse

Connect a USB wired mouse to one of the other larger USB-A sockets on Raspberry Pi. As with the keyboard, it is possible to use a Bluetooth wireless mouse, but setup is much easier with a wired connection.

03 HDMI cable

Next, connect Raspberry Pi to your display using an HDMI cable. This will connect to one of the micro-HDMI sockets on the side of a Raspberry Pi 5 / 4, or full-size HDMI socket on a Raspberry Pi 3/3B+. Connect the other end of the HDMI cable to an HDMI monitor or television.

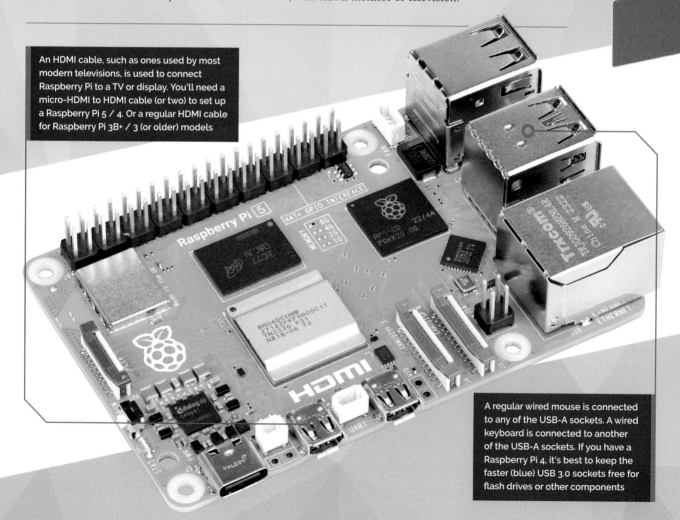

An HDMI cable, such as ones used by most modern televisions, is used to connect Raspberry Pi to a TV or display. You'll need a micro-HDMI to HDMI cable (or two) to set up a Raspberry Pi 5 / 4. Or a regular HDMI cable for Raspberry Pi 3B+ / 3 (or older) models

A regular wired mouse is connected to any of the USB-A sockets. A wired keyboard is connected to another of the USB-A sockets. If you have a Raspberry Pi 4, it's best to keep the faster (blue) USB 3.0 sockets free for flash drives or other components

The USB-C socket is used to connect power to the Raspberry Pi 400. You can use a compatible USB-C power adapter (found on recent mobile phones) or use a bespoke power adapter such as the Raspberry Pi 15.3 W USB-C Power Supply

The Ethernet socket can be used to connect Raspberry Pi 400 directly to a network router (such as a modem/router at home) and get internet access. Alternatively, you can choose a wireless LAN network during the welcome process

SET UP RASPBERRY PI 400

Raspberry Pi 400 has its own keyboard – all you need to connect is the mouse and power

01 Connect a mouse
Connect a wired USB mouse to the white USB connection on the rear of Raspberry Pi 400. The two blue USB 3.0 connectors are faster and best reserved for external drives and other equipment that need the speed.

02 Attach the micro-HDMI cable
Next, connect a micro-HDMI cable to one of the micro-HDMI sockets on the rear of Raspberry Pi 400. The one next to the microSD card slot is the first one, but either connection should work. Connect the other end of the HDMI cable to an HDMI monitor or television.

03 The microSD
If you purchased a Raspberry Pi 400 Personal Computer Kit, the microSD card will come with Raspberry Pi OS pre-installed. All you need to do is connect the power and follow the welcome instructions. If you do not have a Raspberry Pi OS pre-installed microSD card, follow the instructions later in 'Set up the software'.

You'll need this micro USB to USB-A adapter to connect wired USB devices such as a mouse and keyboard to your Raspberry Pi Zero W

Raspberry Pi Zero W features a mini-HDMI socket. You'll need a mini-HDMI to full-sized HDMI adapter like this to connect your Raspberry Pi Zero W to an HDMI display

SET UP RASPBERRY PI ZERO

You'll need a couple of adapters to set up Raspberry Pi Zero and Zero 2

01 Get it connected

If you're setting up a smaller Raspberry Pi Zero, you'll need to use a micro USB to USB-A adapter cable to connect the keyboard to the smaller connection on the board. Raspberry Pi Zero models only have a single micro USB port for connecting devices, which means you'll need to either get a small USB hub or use an all-in-one mouse and keyboard.

02 Mouse and keyboard

You can either connect your mouse to a USB socket on your keyboard (if one is available), then connect the keyboard to the micro USB socket (via the micro USB to USB-A adapter). Or, you can attach a USB hub to the micro USB to USB-A adapter.

03 More connections

Now connect your full-sized HDMI cable to the mini-HDMI to HDMI adapter, and plug the adapter into the mini-HDMI port in the middle of your Raspberry Pi Zero. Connect the other end of the HDMI cable to an HDMI monitor or television.

First, insert your microSD card into Raspberry Pi

With the microSD card fully inserted, connect your power supply cable to Raspberry Pi. A red light will appear on the board to indicate the presence of power

SET UP
THE SOFTWARE

Use Imager to install Raspberry Pi OS on your microSD card and start your Raspberry Pi

Now you've got all the pieces together, it's time to install an operating system on your Raspberry Pi so you can start using it. Raspberry Pi OS is the official software for Raspberry Pi, and the easiest way to set it up on your Raspberry Pi is to use Raspberry Pi Imager. See the 'You'll Need' box and get your kit together.

01 Download Imager
Raspberry Pi Imager is available for Windows, Mac, and Linux computers. You can also install it on Raspberry Pi computers, to make more microSD cards once you are up-and-running. Open a web browser on your computer and visit **magpi.cc/imager**. Once installed, open Imager and plug your microSD card into your computer.

02 Choose OS
Click on 'Choose Device' in Raspberry Pi Imager and select the computer that matches your hardware. Now, click 'Choose OS' and pick the 'Recommended' version of Raspberry Pi OS (typically 64-bit). Click 'Choose Storage' and choose the microSD card you just inserted (it should say 8GB or the size of the card next to it). Click on 'Write'. Your computer will take a few minutes to download the files, copy them, and verify the data.

03 Set up Raspberry Pi
When Raspberry Pi Imager has finished verifying the software, you will get a notification window. Remove the microSD card and put it in your Raspberry Pi. Plug in your Raspberry Pi power supply and, after a few seconds, a blue screen will appear with 'Resizing Filesystem'. It will quickly vanish and be replaced by 'Welcome to Raspberry Pi'. Follow the on-screen instructions to set up Raspberry Pi OS.

Top Tip

Is your card ready?

You don't need to do this if your Raspberry Pi came with a card pre-written with Raspberry Pi OS.

You'll Need

> A Windows/Linux PC or Apple Mac computer

> A microSD card (8GB or larger)

> A microSD to USB adapter (or a microSD to SD adapter and SD card slot on your computer)

> Raspberry Pi Imager **magpi.cc/imager**

Emulate EVERYTHING

Use ready-made emulation distros to turn Raspberry Pi into an all-in-one emulator to play the best classic and modern retro games

BY KG ORPHANIDES

Raspberry Pi can be fully transformed into a DIY multi-console, thanks to three polished emulation-oriented operating systems: RetroPie, Recalbox, and Lakka. They're all excellent and include a wide range of emulators, a notable number of which are maintained by the Libretro project.

All three are available as image files that you can both download and write to an SD card using Raspberry Pi Imager (**magpi.cc/imager**). Once

you've connected them to your network, you can just browse to their SMB shared directories from any computer, and drop your ROM and BIOS files over to play.

CHOOSING YOUR PERFECT DISTRO

All three retro gaming distributions support a wide range of controllers, including USB and Bluetooth console pads, retro joysticks, and full arcade setups connected via USB or GPIO. You can use a keyboard as well, although this is not the best way to navigate their menus and play games, so we strongly recommend investing in a joypad. The Sony DualShock 4 is a reliable choice.

Recalbox has also recently started manufacturing hardware, such as a SCART adapter for Raspberry Pi, making it significantly easier to connect your emulation console to an old-school CRT TV with the Recalbox RGB Dual (**magpi.cc/rgbdua**).

Both Recalbox and RetroPie use the EmulationStation front-end, customised to provide characteristic user experiences, and

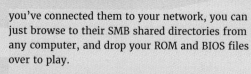

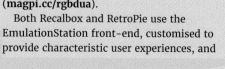

LAKKA

Developed by the Libretro team behind the RetroArch front-end and libretro library of emulator cores, Lakka is a lightweight emulation distro based on LibreELEC. While its interface doesn't feel as polished or attractive as its rivals', Lakka gets the latest stable libretro emulator updates first and releases special versions for a range of emulation hardware, such as DIY handhelds. The other emulation distros we've looked at are often partly built upon Libretro's work. However, while Libretro's arcade and console emulators are among the best around, its vintage microcomputer emulators leave something to be desired, particularly when it comes to keyboard input handling.

lakka.tv

with a range of additional supported themes. If you're building an arcade machine, Recalbox's large, friendly default interface and browser-accessible configuration and game ROM upload tools, alongside SMB shares, make it particularly appropriate for stand-alone cabinets. Recalbox also provides customised support for a number of Raspberry Pi-powered handhelds.

However, Recalbox has a fixed, somewhat limited selection of emulators and versions. These are tried and tested, but it means edge cases won't be catered to when it comes to game support.

Meanwhile, RetroPie's more intimate interface, smaller fonts, and in-depth, exclusively keyboard-navigated emulator module configuration make it a good call for multi-console or even computer emulation, particularly if you've got a keyboard immediately to hand.

RetroPie, built on top of Raspberry Pi OS, is also more customisable and provides a wider range of emulators, including dozens of experimental and niche options. However, unlike Recalbox and Lakka, a 64-bit version of RetroPie isn't currently available. This is largely inconsequential, but a handful of emulators have better performance in their 64-bit versions

Finally, Lakka is the official distro of the RetroArch front-end and libretro emulator ecosystem. Its interface is more workmanlike and bare-bones than its counterparts. This makes Lakka less pick-up-and-play friendly than either RetroPie or Recalbox. However, Lakka makes it very easy to try different emulators on the same game to find the one that works best for it, which has advantages.

> Lakka makes it very easy to try different emulators on the same game

RECALBOX

Recalbox's attractive and easy to navigate EmulationStation-based front-end design works particularly well on handhelds and arcade machines, and custom versions are available for a range of retro console hardware kits, many of which are autodetected in order to load appropriate controller configurations. Recalbox also makes the RGB Dual HAT, custom hardware that adds high-quality CRT display output to Raspberry Pi, and that's only fully supported in this distro, so far. The disadvantage is that Recalbox ships with a fixed range of emulators that you can't add to, upgrade, or downgrade independently of the OS. However, scores of different emulators are included, including non-default alternatives that you can manually select to try specific games with.

recalbox.com

BIOS & ROM LEGAL STATUS

Some emulators require BIOS ROM images taken from the original hardware. These can sometimes be used legally in the UK, because permission has specifically been given by the rights holder to either copy the BIOS onto a modern PC, or download the BIOS ROM from the internet. This table shows all systems that we know either don't require BIOS images, or for which appropriate images are legally available.

PLATFORM	ADDITIONAL BIOS OR FIRMWARE IMAGES REQUIRED?	LEGAL IMAGES AVAILABLE FOR FREE?	SOURCE	LEGAL IMAGES AVAILABLE TO BUY?	SOURCE
3DO	Yes	No	N/A	No	N/A
Amstrad microcomputers	Yes	Yes (since 1999)	magpi.cc/amstradromstxt	No	N/A
Apple II	No	N/A	N/A	No	N/A
Apple Macintosh II	Yes	No	N/A	No	N/A
Atari Jaguar	No	No	N/A	No	N/A
Atari Lynx	Yes	No	N/A	No	N/A
Atari ST	Yes	EmuTOS	magpi.cc/emutos	No	N/A
Commodore Amiga	Yes	AROS open source ROM	aros.org	Yes, from Cloanto	amigaforever.com
Commodore C64	Yes	Yes, from Cloanto	magpi.cc/c64foreverexpress	Yes, from Cloanto	magpi.cc/c64foreverplus
DOS	No	FreeDOS is open source	freedos.org	N/A	N/A
MSX microcomputers	Yes	C-BIOS	cbios.sourceforge.net	No	N/A
Nintendo 64	Optional	No	N/A	No	N/A
Nintendo Entertainment System	No	No	N/A	No	N/A
Nintendo Game Boy	No	No	N/A	No	N/A
Nintendo Game Boy Advance	Optional – required for Joybus link functionality	mGBA HLE BIOS	mgba.io	No	N/A
Nintendo Game Boy Color	No	No	N/A	No	N/A
Nintendo GameCube	Optional – required for some font support	Dolphin HLE BIOS	magpi.cc/dolphinemu	No	N/A
Nintendo SNES	No	N/A	N/A	No	N/A
Nintendo Wii	Yes	Dolphin HLE BIOS	magpi.cc/dolphinemu	No	N/A
Sega Dreamcast	Optional	Flycast/Redream HLE BIOS	magpi.cc/flycast	No	N/A
Sega Game Gear	No	N/A	N/A	No	N/A
Sega Master System	No	N/A	N/A	No	N/A
Sega Mega CD	Yes	No	N/A	No	N/A
Sega Mega Drive	No	N/A	N/A	No	N/A
Sega Saturn	Optional – original BIOS required for multi-disc games	Yabause HLE BIOS	magpi.cc/yabause	No	N/A
SNK Neo-Geo	Yes	No	N/A	Yes	DotEmu SNK 30th Anniversary edition
Sony Playstation	Yes	PCSX-Reloaded HLE BIOS	magpi.cc/pcsx (included in PCSX-Reloaded and PCSX-Rearmed emulators)	Yes	Use RPCS to install PS3 firmware magpi.cc/pssystemsoftware
ZX Spectrum	Yes	Yes (since 1999)	Generally included with emulators	No	N/A

Emulate on the
DESKTOP

Build an old-school micro or desktop PC gaming experience on top of Raspberry Pi OS

Like all UAE derivatives, AmiBerry's settings interface can feel overwhelming, but you'll generally be OK using just the Quickstart, Input, and Disk swapper settings. Buying original system ROMs is strongly recommended, though

W hile having a DIY games console, handheld, or arcade machine is incredibly cool, it's not always the best setup for playing old school computer games, especially those that require a keyboard and mouse. For this, a proper desktop environment really comes into its own, although you may still want to reduce your resolution to optimise the performance and appearance of older titles.

DOS and early Windows emulation are probably the most popular and accessible options, but Microsoft operating systems and Intel hardware aren't the only computer platforms that lend themselves to easy desktop emulation. We'll also run C64, Commodore Amiga, and Atari ST games.

You can configure Amiberry to use familiar WASD keyboard controls in the place of a joystick, but you're still better off connecting a modern joypad

01 GET READY TO PLAY

Create directories in your home directory called **Games** and **Software**. Some of our emulators will live in **Software**, but let's grab some games first. Create subdirectories for each system you're going to emulate to house the games we'll be playing: **atari**, **amiga**, **c64**, and **dos**. After downloading them, extract each of the compressed game files into an appropriately named subdirectory of those folders.

02 DOWNLOAD FREE GAMES

A number of developers have made freeware versions of their games available at **scummvm.org/games**. There are plenty to choose from, but we'll use the CD version of Revolution Software's Beneath a Steel Sky and extract it into **~/Games/dos/BASS**. Sticking with the classics, grab the shareware version of Doom 1.9 for DOS from **magpi.cc/doom19** and unzip it into **~/Games/dos/doom**.

YOU'LL NEED

> Raspberry Pi
> Raspberry Pi 64-bit OS
> Optional joystick or joypad (recommended for ST and Amiga games)

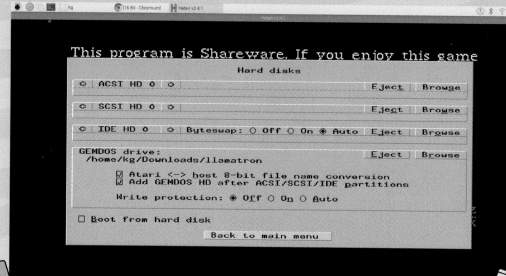

Hatari is one of the most faithful ST/STe emulators around, but you can still just mount a local directory full of PRG files as a virtual hard disk. It's not immediately obvious

On the Atari ST front, Jeff Minter's glorious Llamatron is available at **minotaurproject. co.uk/lc-16bit**. For the Amiga, you can download Sword of Sodan from its creator's site at **magpi.cc/sodan**. Our C64 game is Nixy the Glade Sprite, available from **magpi.cc/nixy** on a pay-what-you-want basis, starting at free - create a subdirectory called **nixy** under **c64**.

03 INSTALL THE LATEST VERSION OF SCUMMVM

To get the latest version of ScummVM with support for the widest range of games, go to **scummvm.org/downloads** and download the appropriate 32-bit (armhf) or 64-bit (aarch64) AppImage for Raspberry Pi OS. At the time of writing, the current release version is 2.7.0. Put the file wherever you like - perhaps in a **Software** directory. Right-click it, select Properties, go to

ScummVM makes it easy to experience and save-scum around the regular deaths that embody vintage adventure games

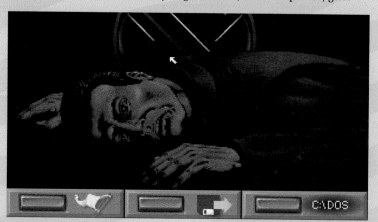

the Permissions tab, open the Execute pulldown and select Only Owner, then click OK. Double-click on the file and select Execute when prompted. Assuming you want useful shortcuts, click Yes when a prompt asks you if you've like to integrate ScummVM with your system.

04 PLAY ADVENTURE GAMES

Once installed, you'll find ScummVM in the Games section of the Raspberry Pi OS desktop menu after this. To add a game, click Add Game..., browse to the directory you unzipped Beneath a Steel Sky into, and select Choose before following the prompts. You can also use the 'Mass add' option available as a pull-down to point ScummVM at your games directory.

05 ADD FLATPAK SUPPORT

The easiest way to install and update the 64-bit version of DOSBox-X for Raspberry Pi is via Flatpak, so we'll start by adding support for this convenient package management system, which bundles up all of a program's dependencies. Open a terminal and type:

```
sudo apt install flatpak
```

Log out of your desktop session and log back in again to ensure that Flatpak's environment variables are now in the path. Open a terminal and type:

```
flatpak remote-add --if-not-exists flathub
https://flathub.org/repo/flathub.flatpakrepo
```

06 INSTALL DOSBOX-X AND DOOM

In a terminal, type:

```
flatpak install dosbox-x
```

Say Yes to all options to install it. Reboot and you'll find DOSBox-X in your Games menu. The first time you run DOSBox-X, you'll be prompted to select a working directory. Select the **~/Games/dos** directory we made earlier and click OK. At DOSBox's command prompt, type:

```
mount c ~Games/dos
c:
cd doom
install
```

If you follow the suggested defaults, once installation is complete, you'll be prompted to launch the game, which will be installed to **~/Games/dos/DOOMS**.

07 INSTALL HATARI

Like DOSBox-X, the best way to keep up-to-date with the Hatari Atari ST and STE emulator on Raspberry Pi running a 64-bit operating system is to install it as a Flatpak. Open a terminal and type:

```
flatpak install hatari
```

Then answer **Y** to all prompts to complete installation. As Flatpak is depreciating 32-bit

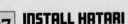

packages, those running 32-bit operating systems on Raspberry Pi will have to use the version in the OS's Apt repository. This is a little older but should still do everything you need it to. Open a terminal and type:

```
sudo apt install hatari
```

Press **F12** (**Super** + **F2** on a standard Raspberry Pi keyboard) to access settings.

08 PLAY ST GAMES

Hatari can be a little awkward when it comes to browsing your entire home directory. The Flatpak installation will set your home directory as its default path, but during our last test, only files and subdirectories in our **Downloads** directory were shown, and you can't type in the path of your choice. If you experience this issue, place

> On the Atari ST, Jeff Minter's glorious Llamatron is available at the Minotaur Project

the directories containing your Atari ST games into **Downloads**.

If you've got ST files, these are usually floppy disk images, and should be mounted from floppy disks. If you've got PRG executables, like our copy of Llamatron, you'll need to mount it as a virtual hard disk. Open settings, click the Hard disks button and, under GEMDOS drive, browse to the folder containing your PRG files and select it.

Tick both the 'Atari <> host 8-bit file name conversion' and 'Add GEMDOS HD after ASCI/SCSI/IDE partitions' boxes, then go back to the main menu.

Click 'Joysticks setup' to check your joypad configuration. If you've not got one, select 'use keyboard' - this defaults to the arrow keys to move and right control to dire, but if you prefer WASD, just click 'Define keys'. Go back to the main menu, select 'Reset machine' and click OK.

QUICK TIP

Courier DOSBox-X

Flatpak images are 64-bit only. If you're running a 32-bit distro, refer to our guide at **magpi.cc/builddos**.

◄ While many developers have sold rights to their games to re-release houses. Llamasoft has made multi-platform classics free to download

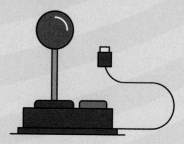

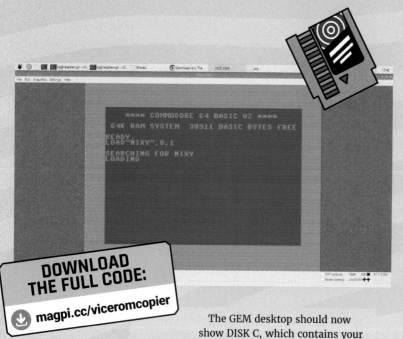

▲ Vice's loading times can be authentically slow, but you can hit **ALT + W** to enable Warp Mode. Just remember to turn it off before you start playing

QUICK TIP

More than just point and click

ScummVM started with adventure games, but now plays many classic RPGs, too. See its full supported game list at magpi.cc/scummvmsupport.

The GEM desktop should now show DISK C, which contains your programs. Double-click to open it and then run the **TRON_512.PRG** file.

09 INSTALL AMIBERRY

Download the latest version of Amiberry for your Raspberry Pi from **magpi.cc/amiberry**. Create an Amiberry subdirectory under **~/Software** and extract the compressed file into that. For the most reliable loading of games from ADF disk images, including Sword of Sodan, you should ideally buy the original Amiga Kickstart ROMs from Amiga Forever (**amigaforever.com**) and place them in **~/Software/Amiberry/kickstarts**. Now open a terminal and type:

```
sudo apt install libsdl2-2.0-0 libsdl2-
ttf-2.0-0 libsdl2-image-2.0-0 flac mpg123
libmpeg2-4 libserialport0
cd Software/Amiberry
```

10 PLAY SWORD OF SODAN

In a terminal, enter:

```
./amiberry
```

It'll open on the Quickstart screen. Tick the boxes next to DF2 and DF3 to enable extra virtual floppy drives. Now click on the Disk swapper tab and, for each of the first four slots, browse to one of the four ADF files (sodan1 through 4) in your extracted copy of Sword of Sodan. If there's a blank space instead of the drive number next to the file path, click on the button – these should say DF0: through DF3:. Finally, configure your controls. Go to input and, if you're not using a

joystick, enable the Keyboard Layout of your choice for Port 1. Finally, click Start. When using your emulated Amiga, press **F12** at any point to open Amiberry's settings. To wield your sword, press fire at the same time as the direction in which you wish to attack.

11 INSTALL VICE

The VICE Commodore 8-bit emulator has returned to Raspberry Pi OS's repositories, now unencumbered by ROMs with potentially disputable licences. That means you'll have to get hold of your own C64 ROMs. Fortunately, they're available for free from Cloanto. This is distributed as a Windows MSI file, so we'll use the Gnome MSI Tools to extract the files it contains and we've written a script to move them to the right places.

Go to **c64forever.com** and click Download Free Express Edition. Open a terminal and type:

```
cd Downloads
sudo apt install msitools vice
mkdir C64Forever && mv C64Forever10.msi ~/
Downloads/C64Forever && cd C64Forever
msiextract C64Forever10.msi
cd 'Program Files/Cloanto/C64 Forever/Shared'
wget https://codeberg.org/MightyOwlbear/VICE-
Forever-ROM-copier/raw/branch/main/copyrom.sh
sudo sh copyrom.sh
```

12 RUN A C64 GAME

Make sure your controller is plugged in, then go to the terminal and type:

```
x64
```

Now you've run VICE's C64 emulator, let's configure it and give it something to emulate. Go to Settings > Settings and select Input Devices > Joystick. Assign your controller to Joystick #2.

Open the File menu and select Smart Attach Disk/Tape/Cartridge.

Browse to **~/Games/c64/nixy/prg** and select **nixy.prg**. Once it's loaded – this takes a minute or so – READY will be displayed on the emulated computer. Now type RUN.

Press **ALT + D** for full-screen mode. If your sound isn't working, try pressing **ALT + W** to disable warp mode, which gives you faster load times but automatically disables audio.

Software HUNTING

▲ The DotEmu ports of SNK's classic beat-'em-ups and shooters like Ironclad include everything you need to run them on the emulator of your choice

The MagPi magazine's guide to buying the best retro re-releases and modern classics

Retro DOS and early Windows games are both popular and easily available, thanks to official re-releases on digital store-fronts like GOG, Zoom Platform, and even Steam, which kept the games accessible, even after the original boxed releases became collectable and started commanding wildly excessive prices on eBay. Meanwhile, homebrew scenes have kept alive classic 8-bit and 16-bit consoles and computers, with varying levels of support from their original creators.

This emulation boom has in turn prompted some new publishers to buy up the rights to ZX Spectrum, C64, Amiga and Atari ST classics and sell them for anywhere from £3 to £25 a pop. Helpfully, almost all of these include original disk images that you can run on any emulation system, but watch out for remakes that only work on modern computes. But vintage games often pale in comparison to the quality and innovation of modern commercial and homebrew games.

BUYING DOS GAMES

GOG usually packages its retro games for Windows users, often bundled with DOSBox or ScummVM, using InnoSetup. Using the innoextract tool from Raspberry Pi OS's repositories, we can extract those game files to use with our own emulator without having to install them on a separate PC with x64 or x86 architecture. You'll even find several vintage classics available for free at **magpi.cc/gogfree**.

BUYING RE-RELEASES ON STEAM

A number of publishers are putting out emulated retro games on Steam. These include Pixel Games (**magpi.cc/pixelgames**), SNEG (**magpi.cc/sneg**),

whose releases you'll also find on GOG, the often-resurrected Apogee Entertainment (**magpi.cc/apogee**), the even more frequently resurrected Atari (**magpi.cc/atari**) - which also publishes some new releases, and Ziggurat Interactive (**magpi.cc/ziggurat**), among others.

SEGA CLASSICS

Sega is famously generous with its Mega Drive images. Just buy a copy of Sega Mega Drive & Genesis Classics on Steam (or any other digital store-front), go to the folder called 'uncompressed ROMs' and copy them over to your emulator. You may need to rename some extensions to .bin, depending on which emulator you use.

SNK

Some DRM-free versions of Dotemu's 25th and 30th anniversary SNK arcade re-releases include a game ROM image and a file called neogeo.zip, which contains SNK arcade machine BIOS images, but not every version includes the full set. The best place to get these is from GOG's stand-alone SNK titles, such as The Last Blade (**magpi.cc/thelastblade**). Run the Linux installer on an x86 PC, as the Windows GOG version no longer includes the BIOS file.

Note that the 40th anniversary releases produced by Digital Eclipse and released on GOG and Steam 40th are not so helpful, and instead contain everything in .mbundle files which require an inconvenient and legally dubious amount of unpacking to actually spit out the game and BIOS files. *M*

▲ A number of retro computer games have made their way onto Steam, such as Gremlin Graphics' sci-fi platform shooter, Venus the Flytrap

▼ Modern C64 games like Nixy the Glade Sprite more than stand up to what developers were creating in the 1980s

NEXT GEN RETRO

on Raspberry Pi 5

FEATURES LEGAL ROMS!

Smoothly emulate fifth- and sixth-generation consoles, including the original PlayStation, Dreamcast, Game Boy Advance and Wii

By KG Orphanides

Warning!
Copyright!

Raspberry Pi 5 is significantly more powerful than its predecessors, and that translates to upgraded emulation performance. While Raspberry Pi 4 could emulate fourth-generation consoles like the Sega Mega Drive and Super Nintendo without breaking a sweat, and could even manage fifth-generation PlayStation games at CRT resolutions, Raspberry Pi 5 opens the door to emulating fifth- and sixth-generation console hardware, like the Dreamcast, GameCube and Wii.

When it comes to desktop emulation, there are a couple of things to be aware of. Raspberry Pi 5 is supported by a new version of Debian-based Raspberry Pi OS, Bookworm. This new OS has moved from using the X Window system (Xorg, X11, or just X) to Wayfire as the default

compositor – the display system that renders your graphical desktop for you. Wayfire uses the efficient new Wayland protocol, developed by **freedesktop.org**, the same project behind Xorg. (See "Where there's a Wayland" boxout).

At the time of writing, the mesa 23.3 graphics stack, although released, hadn't yet made it into the Raspberry Pi OS repositories, but should result in improved Vulkan support on Raspberry Pi 5.

YOU'LL NEED

- Raspberry Pi 5
- USB or Bluetooth gaming controller (such as 8BitDo Pro, **magpi.cc/8bitdo**)
- Raspberry Pi OS Bookworm
- An x86 computer (optional). Required to obtain the official PlayStation BIOS image

Raspberry Pi 5's faster hardware has opened up a whole new range of emulatable consoles

You'll want a controller, like this 3D-printed Alpakka (**magpi.cc/alpakka**), to play retro games

WHERE THERE'S A WAYLAND

Wayland is less mature than X11, and some software hasn't been updated to work with it, although Raspberry Pi OS does ship with XWayland, a compatibility tool that allows you to run individual programs in X11, which it'll automatically fall back to. You can also use the WAYLAND_DISPLAY="" <your-app-name> command to force its use.

It's possible, via raspi-config, to switch Bookworm over to the Openfire window manager with a legacy X11 backend, used by Raspberry Pi 3 and earlier systems, but we're going to stick with Wayland as this is where most active development is happening.

Raspberry Pi OS is also moving away from the OpenGL graphics rendering API to Vulkan. Legacy support is built in and most programs will fail over to either software rendering or OpenGL.

EMULATING WII AND GAMECUBE CONSOLES

▼ There's some fantastic Wii homebrew out there, like synthwave racer *Newo Zero*

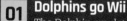

Top Tip

Build your own GameCube game

Retro League, a car football game for GameCube using a novel game engine, is open source and ready to build or modify (on x86 architecture) at **magpi.cc/retroleague**.

01 Dolphins go Wii

The Dolphin emulator allows you to run GameCube and Wii games. Sadly, you can't directly play Nintendo optical discs in Dolphin and it's illegal to copy console game discs in the UK, but there's a vibrant Wii homebrew scene. Although there's a copy of dolphin-emu in the Raspberry Pi OS repositories, there's a newer one available, so we're going to build from source. Open terminal and enter the following:

```
$ sudo apt install ca-certificates qt6-base-
dev qt6-base-private-dev libqt6svg6-dev git
cmake make gcc g++ pkg-config libavcodec-dev
libavformat-dev libavutil-dev libswscale-dev
libxi-dev libxrandr-dev libudev-dev libevdev-
dev libsfml-dev libminiupnpc-dev libmbedtls-dev
libcurl4-openssl-dev libhidapi-dev libsystemd-
dev libbluetooth-dev libasound2-dev libpulse-
dev libpugixml-dev libbz2-dev libzstd-dev
liblzo2-dev libpng-dev libusb-1.0-0-dev gettext
$ git clone https://github.com/dolphin-emu/
dolphin.git
$ cd dolphin
$ git submodule update --init --recursive
$ mkdir Build && cd Build
$ cmake .. -DUSE_SHARED_ENET=ON
$ make -j$(nproc)
$ sudo make install
```

Now enter:

```
$ dolphin-emu
```

...to start the Dolphin emulation software.

02 Connect a controller

If you're using a USB controller, plug it in. If you're using a Bluetooth controller, open Raspberry Pi's Bluetooth menu towards the top right of the desktop taskbar and click Add Device. Put your controller in pairing mode and select it from the Add New Device menu when it appears, most likely as "Wireless Controller". We used a Sony DualShock 4 controller, but you can use any USB or Bluetooth controller with Dolphin, including original Wii wireless controllers. We'll be using the same controller with all our emulators.

03 Configure a GameCube controller

In Dolphin, open the Options menu and select Controller Settings. Here, ensure that the Port 1 GameCube controller is set to Standard Controllers, Wii Remote 1 is set to Emulated Wii Remote, and make sure Background Input is ticked at the bottom. Click the Configure button for the GameCube controller. In the GameCube Controller at Port 1 window, under Device, select your controller, then calibrate each button to your preferred layout by clicking on the button assignment and then pressing the controller button or movement you want to associate it with. The Modifier for each thumbstick should generally be set to that stick's click. The GameCube controller doesn't map entirely cleanly to most modern controllers. The Z button should be mapped to your controller's right shoulder buttons. This diagram will help with mapping: **magpi.cc/ gcconwiki**. Give the profile a name and click Save before closing the window.

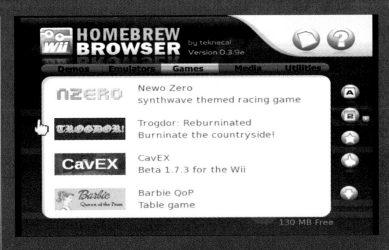

04 Configure a Wii controller

Next, we'll do the same mapping for an emulated Wii remote. This is an even more unusual controller, so we're just going to have to do our best here. At least there aren't many buttons to map. We mapped the D-pad to our D-pad, AB12 to our front buttons, and + and - to the left and right shoulder buttons. Open the motion emulation tab and map pointer controls to your left analogue stick. Name your file, save and then close the configuration window.

> " **There's a vibrant Wii homebrew scene** "

05 Getting started with Wii homebrew

There are plenty of homebrew games for the Wii but getting them running on Dolphin can be fiddly. Using the Open Shop Channel (**oscwii.org**) to install homebrew gives you a much higher success rate.

The OSC provides a similar experience to the software download channels that Nintendo used to offer for the Wii. You'll need a virtual SD card to be able to download games from the channel, but Dolphin will create this for you.

First, we'll create a directory where you'll store your games and download the open homebrew browser. Open a terminal and type:

```
$ mkdir -p Games/Wii
$ cd Games/Wii
$ wget https://hbb1.oscwii.org/unzipped_apps/
homebrew_browser/apps/homebrew_browser/boot.dol
```

06 Access the Open Shop Channel

In Dolphin, double-click the main pane to select your games directory – browse to the Games/ Wii folder we just made. Click the Config button on the toolbar, select the Paths tab, and tick the Search Subfolders box. With that done, a file called `bool.dol` should now appear on Dolphin's main screen.

Double click it to launch the browser and, if this is your first launch, give it a minute to attempt to contact the server for updates. Press B to skip the update. You can press A to install it, which inevitably fails, as there's not been a new version released since 2012, although new software has been added to the Open Shop Channel itself.

Either way, the Homebrew Browser will then open. You'll find numerous freeware and open-source games and programs here. Try *Newo Zero*, a chill synthwave-themed racing game, and music utility *Harmonium*.

▲ The best way to access Wii homebrew is to install the Homebrew Browser and access the Wii Open Shop Channel

07 Configure your storage

In Dolphin, open the Options menu and select Configuration. Select the Wii tab. Under SD card settings, you'll see that Insert SD card is already enabled. Below it, tick Automatically Sync with Folder – this will give us an editable version of the SD card that we can browse and modify as needed, and will sync changes every time you run Dolphin.

Dolphin automatically creates a sync folder in your `~/.local/share` tree, but this won't be indexed by default. Instead, create a WiiSD folder in the `~/Games/Wii` directory we made earlier and select it.

Click Convert File to Folder Now, and the games you downloaded to your SD card in the previous step will appear in Dolphin's main game menu, easily available for you to run. Not every homebrew game runs smoothly under emulation, but you'll have fun finding those that do.

EMULATING GAME BOY ADVANCE

01 GameCube's pocket pal

While the Wii is well supported, you'd struggle to find GameCube homebrew titles. Released in the same year as the GameCube, the Game Boy Advance was its more successful, pocket-sized sibling, and has one of the most vibrant dev communities around. We're going to install the mGBA emulator as a Flatpak package, so we'll start by installing the package manager and repository. The Flathub app store (**flathub.org**) only supports 64-bit operating systems and processors, tagged aarch64 in its index. Open a terminal and type:

```
$ sudo apt install flatpak
$ flatpak remote-add --if-not-exists flathub
https://flathub.org/repo/flathub.flatpakrep
```

Now reboot Raspberry Pi to ensure that Flatpak's desktop integrations can set themselves up correctly. With that done, you'll be able to search for software from the command line or at **flathub.org**, although Flatpak installation from the browser isn't currently supported by the Raspberry Pi OS software centre.

▼ With a Game Boy Advance emulator like mGBA, you can buy and run new digital releases like the excellent adventure platformer *Goodboy Galaxy*Channel

message telling you that it's failed back to software rendering mode and Wayland won't correctly render the window, although the emulator will be usable. This is likely to be resolved when Mesa 23.3 comes to Raspberry Pi.

Until that's fixed, we'll run mGBA with a terminal command:

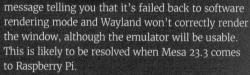

```
WAYLAND_DISPLAY="" flatpak run io.mgba.mGBA
```

We're still on software rendering, but now the mGBA desktop application can be moved and resized.

Beyond these teething problems, mGBA works out of the box. Just load a ROM in .gba format from the File menu. You can play in fullscreen mode by pressing **CTRL+F**.

02 GBA: Online

Open a terminal window and type: `flatpak install flathub io.mgba.mGBA.` Choose "yes" when prompted to confirm the package and its dependencies, and it'll be added to your Games menu. If you open it from there, though, you'll get an error

03 Get some GBA games

There's been a flurry of recent GBA releases, largely distributed on the itch.io digital store. We've been playing the fantastic *Goodboy Galaxy* (**magpi.cc/goodboygalaxy**), and you can download the prologue/demo for free at **magpi.cc/goodboygalaxydemo**. Help a cat rescue their family in free platformer *Feline* (**magpi.cc/feline**), tower-defend your oozy kingdom in *Slimelord Siege* (**magpi.cc/slimelordseige**) or go dungeon delving in *Inheritors of the Oubliette* (**magpi.cc/oubliette**).

You can check out over 100 more GBA games in our Itch collection at **magpi.cc/itchgba**.

EMULATING DREAMCAST

01 Dream on

We'll install the Flycast emulator to play Dreamcast games. Flycast is also available as a flatpak, but at the time of writing this wasn't working reliably on Raspberry Pi OS. This is likely to be fixed by the time you read this. Open a terminal and type:

```
$ git clone https://github.com/flyinghead/
flycast.git
$ cd flycast
$ git submodule update --init --recursive
$ mkdir build && cd build
$ cmake ..
$ make -j4
$ sudo ln -s ~/Software/flycast/build/flycast
/usr/local/bin
$ flycast
```

02 Configure Flycast

Click the Settings button at the top right of the Flycast window and, in the General tab, click the add button in the Content Location bar. This should direct you to the directory where you'll be keeping your Dreamcast games. We've opted to put them in ~/Games/Dreamcast. Put each game in its own directory – Flycast will check all subdirectories automatically. Now click on the Advanced tab, and under Other, tick the HLE BIOS box. You can optionally keep each game's save files on its own virtual VMU (Visual Memory Unit, also known as a Visual Memory System) by ticking the per-game VMU A1 box at the bottom of the Controls tab. Now click Done.

03 Get some Dreamcast homebrew

Now we'll need some games. As with many classic consoles, hobbyists and indie developers are still making both free and commercial software for the Dreamcast. Crazy Viking Studios granted Marc Hall permission for their 2013 platformer *Volgarr the Viking* to be ported to the Dreamcast and distributed for free. Download a copy from **volgarr.rkd.zone** and unzip the file into a subdirectory of your Dreamcast folder. If you've got Flycast open, the game should immediately appear as a playable disc image.

For a feel of what DC homebrew was like back in the day, download the Inducer collection from **magpi.cc/inducer** and extract it into your Dreamcast games directory, but be aware that the games in this collection don't always emulate smoothly. You'll find more Dreamcast homebrew on the Dreamcast Talk forum's new releases pages (**magpi.cc/dreamcasttalk**) and in our Itch collection **magpi.cc/itchdreamcast**). The Dreamcast Junkyard (**thedreamcastjunkyard.co.uk**) regularly covers new releases.

▼ Arcade platformer *Volgarr the Viking* was ported to Dreamcast in 2015 with the blessing of the original developers

WHAT'S A VMU?

Visual Memory Unit (VMU) devices were the Dreamcast's external flash memory card modules, which plugged into the controllers, allowing players to take their saved data around if they wanted to visit friends. They had mono LCD screens, integrated file managers and the ability to double as a handheld console. A number of Dreamcast games allowed you to load extra mini-games onto the VMU. Flycast doesn't allow you to play these directly, but it can load the Dream Explorer (**magpi.cc/dreamexplorer**) homebrew, which allows you to manage your VMUs.

EMULATING PLAYSTATION

01 PlayStation 1 emulation

PlayStation emulation is mature, but you get different features from different emulators. If you primarily want to play homebrew, then Mednaffe – a front end for the Mednafen multi-emulator - is an excellent choice, as it upscales well. However, it doesn't include an emulated BIOS and can't play games from disc.

There are others, including PCSXR, bus as we're interested in new games, we'll stick with Mednaffe here.

Open a terminal and type:

```
$ sudo apt install mednaffe
```

02 Getting an official PS1 BIOS

You can currently get a PS1 BIOS from Sony, as it's included in a PS3 BIOS that's available. However, you'll need to install the RPCS (**rpcs3.net/download**) emulator to open it, and that's most effectively done on an x86 PC.

Go to **magpi.cc/pssystemsoftware** then click the + sign next to Reinstall using a computer, then click the Download PS3 Update button.

Install and run RPCS. Select Install Firmware from its File menu and select the PS3UPDAT.PUP file you downloaded and allow it to install.

With this done, assuming you're running RPCS3 on Linux, you'll find a file called `ps1_rom.bin in ~/.config/rpcs3/dev_flash/ps1emu/`

More detailed instructions are available at **magpi.cc/ps1bios**. To work with Mednafen, you'll have to rename the file. ps1_rom.bin is a region-free universal ROM, so can be renamed to reflect any of the three PlayStation region. Just copy it

three times, and name the copies scph5500.bin, scph5501.bin and scph5502.bin.

03 PS1 Homebrew

The original Net Yaroze development kit is still the gold standard for homebrew making on PlayStation, and you'll find plenty of games and demos to download at **magpi.cc/netyaroze**, as well as a few at PSX Palace (**magpi.cc/psxpalace**). Download Kaiga's action–RPG *Magic Castle* from magpi.cc/magiccastle – we'll use this in the next step, so unzip it in a suitable directory - ours is in `~/Games/PS1/Magic_Castle.`

04 Configuring Mednaffe for PS1

First, you'll need to copy your renamed BIOS files over to Raspberry Pi and put it into `~/.mednafen/firmware/` - you may need to run Mednaffe once to create this directory. Start Mednaffe from the menu or a terminal window.

Before we play, we'll need to configure a few things:

Select the Global Setting tab, then Sound, and set the driver to sdl. Go to Systems, select the Input, choose Sony PlayStation and then select the Input tab from the pane on the right. Select each original PlayStation controller button and map a keyboard or controller key to it.

Finally, open the File menu > Open, browse to Magic_Castle_2021_07_May.cue and open it. Note that Mednaffe uses bin/cue file pairs rather than iso images. Press **ALT**+**ENTER** to fullscreen your game, and enjoy *Magic Castle*.

Top Tip

PlayStation 2

PlayStation 2 games are playable using the Play! emulator (**purei.org**) but performance on Raspberry Pi 5 is limited for many titles.

▼ Recalbox automatically configures its PlayStation emulator for your controller and gives you a nice on-screen bezel, shown here around free action RPG Magic Castle

EASY PLAYSTATION EMULATION WITH RECALBOX

While running emulators from your desktop gives you more control over your downloads and settings, there's nothing like a dedicated emulation distro for convenience. Recalbox was first off the mark with dedicated support for Raspberry Pi 5. At the time of writing, Recalbox 9.2-Experimental for Raspberry Pi 5 wasn't yet available for direct installation from within Raspberry Pi Imager.

Download the image file manually from **magpi.cc/recallboxrpi5**, open Raspberry Pi Imager, and from "Choose OS", scroll down and select "Use custom". Select the image file you just downloaded and the Storage menu to select your microSD card, click "write", have a cup of tea, then pop the microSD card into Raspberry Pi 5. Recalbox supports external storage, but if you want a self-contained unit, use a large microSD card (over 64GB), as CD games can take up a fair bit of space.

01 Essential files

Press the **START** button on your controller, go to Network Settings, enable Wi-Fi, and enter your Wi-Fi password. When connected to a network, Recalbox presents SMB shares and a web interface (accessible by default via a browser at **http://recalbox.local** if you're on Linux or **http://recalbox/** if you're on Windows). The SMB shares allow you to easily upload games and BIOS images, while the web interface gives you easy remote access to Recalbox's settings and at-a-glance views of installed games and BIOSes. The easiest way to add new BIOS images is to access **smb://recalbox.local/bios/** from another computer and copy files over. Recalbox ships with a number of freeware and emulated BIOS files, and all of the legally available BIOS files for various systems

that we've discussed here and in previous tutorials such as our Emulate Everything guide in issue 133 (**magpi.cc/133**) should also be compatible, making this an excellent all-in-one emulation solution.

02 Add your games collection

Recalbox ships with a variety of open-source and freeware games, and you'll find a wide range of additional platforms ready to go. After copying your game and BIOS ROMs over, you'll want to reboot to force Recalbox to re-scan everything. You can always exit a game by pressing the controller hotkey and **START** buttons.

> **" You get different features from different emulators "**

03 Add more storage

Connect an external drive and boot Recalbox, press **START** and go to System Settings. Go down to the Storage Device entry and select your USB stick. Recalbox will reboot and create a `/recalbox` directory tree on the device for your files. **M**

Top Tip

Music, maestro

If you're using a Sony DualShock 4 controller, it can end up being the default audio device. In Recalbox, fix that in Settings > Sound Settings > Output Device.

▼ You can configure Recalbox to use a USB drive as its main hard disk for game and BIOS storage

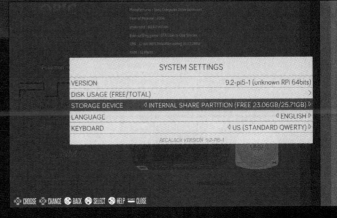

SYSTEM SETTINGS	
VERSION	9.2-pi5-1 (unknown RPi 64bits)
DISK USAGE (FREE/TOTAL)	>
STORAGE DEVICE	◁ INTERNAL SHARE PARTITION (FREE 23.06GB/25.71GB) ▷
LANGUAGE	◁ ENGLISH ▷
KEYBOARD	◁ US (STANDARD QWERTY) ▷
RECALBOX VERSION 9.2-PI5-1	

⊕ CHOOSE ⊕ CHANGE ✕ BACK ✕ SELECT ✕ HELP ═ CLOSE

Play classic console games legally on Raspberry Pi

Discover a range of ways to buy and source classic games legally for Raspberry Pi

MAKER

K.G. Orphanides

K.G. is a writer, maker of odd games and software preservation enthusiast. They will fight anyone who claims that piracy is the only thing emulation's good for.

@KGOrphanides

Console emulation has been firmly in the mainstream in recent years. However, hobbyist emulation and DIY consoles run the risk of involving you with illegal copyrighted content. But you don't have to be a bootlegger to build your own home multi-console emulation with Raspberry Pi and RetroPie.

Emulators themselves are strictly legal, and we've talked in the past about the wide range of homebrew and legal ROM images available (**magpi.cc/legalroms**).

In this tutorial we're going to look at a much broader range of legal console ROMs. Some can be purchased legally, while others have been developed and are distributed for free.

So let's set up a RetroPie console and play some classic games.

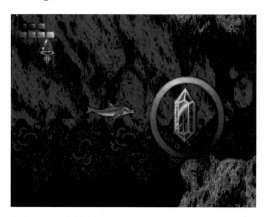

▲ Ecco the Dolphin is just one of the classic games in the Mega Drive Classics collections

Thriving scene

Sega's Mega Drive Classics collections include ROM images of the games that you run on any emulator you like, and brand new commercial releases for Sega and other platforms are thriving, as are active homebrew scenes bringing innovative new games to console formats that went out of production over 25 years ago.

Sega is incredibly supportive of its emulation community, and is happy to just sell you some classic Mega Drive ROMs, included in the Sega Mega Drive Classics collections for Windows, macOS, and Linux. You can buy 50 classic Mega Drive games on Steam (**magpi.cc/segaclassics**), either individually or as a pack.

Once bought, to find the ROMs, open the title's Steam Library page, clear the gear icon on the right, select properties Properties, select the Local files tab, and then click Browse local files. You'll find all the ROMs in the clearly labelled uncompressed ROMs directory. Rename all files with '.68K' and '.SGD' extensions to '.bin' and copy them over to Raspberry Pi using a USB stick or via its Samba share (**magpi.cc/samba**).

Buy new classics

If you're after new games for classic systems, **itch. io** should be your first port of call. The Nintendo Entertainment System is the most popular 8-bit console for modern developers, while the Mega Drive has won the hearts of 16-bit devs. Games are

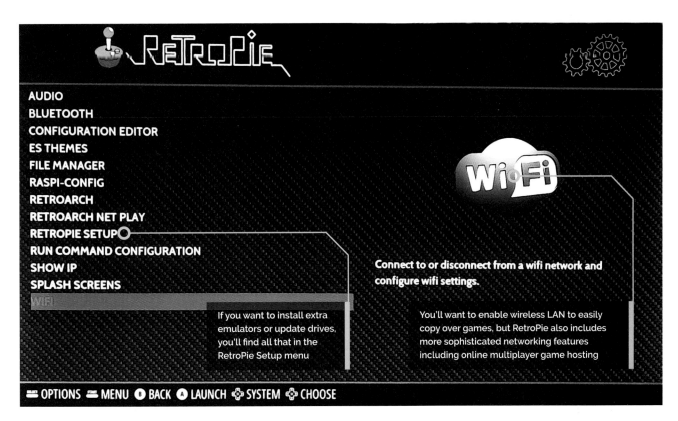

RetroPie

AUDIO
BLUETOOTH
CONFIGURATION EDITOR
ES THEMES
FILE MANAGER
RASPI-CONFIG
RETROARCH
RETROARCH NET PLAY
RETROPIE SETUP
RUN COMMAND CONFIGURATION
SHOW IP
SPLASH SCREENS
WIFI

If you want to install extra emulators or update drives, you'll find all that in the RetroPie Setup menu

Connect to or disconnect from a wifi network and configure wifi settings.

You'll want to enable wireless LAN to easily copy over games, but RetroPie also includes more sophisticated networking features including online multiplayer game hosting

OPTIONS MENU ⓑ BACK ⓐ LAUNCH ✛ SYSTEM ✛ CHOOSE

---

also available for the 8-bit Sega Master System and 16-bit Super Nintendo Entertainment System.

We've made itch.io collections for each of those platforms, going out of our way to avoid unauthorised ports and ROM hacks. These include both commercial and freeware games, plus a couple of open-source titles.

That's not the only place that you'll find releases for those platforms. In the tutorial, we download Mystery World Dizzy by the Oliver Twins. It's a wonderful example of a lost NES game that was recovered by its creators and released as freeware to the fan community, but it's also rare to find Nintendo games from that era re-released with their creators' blessing. And unlike Sega, Nintendo doesn't look fondly on ROM hacks, fan games, and the like.

On the homebrew side of things, projects such as Retrobrews (**retrobrews.github.io**) and sites like **vintageisthenewold.com** and **indieretronews.com** compile collections and lists of homemade games for classic consoles, but watch out for the odd unauthorised port slipping into their catalogues.

There's a small but lively industry releasing cartridges for retro consoles, and a number of developers and publishers make the ROM files available online, either for free or a small price. Among these are RetroSouls (**retrosouls.net**), the team behind Old Towers, and Miniplanets publisher Playonretro (**playonretro.itch.io**).

If you buy Sega's Mega Drive Classics collection on Steam, you'll get emulator-friendly ROM files for 50 games, including Golden Axe, Ecco the Dolphin, and Sonic the Hedgehog

> **Sega is incredibly supportive of its emulation community**

Eight modern Mega Drive games

Here are some of the best new Mega Drive games:

Tanglewood – magpi.cc/tanglewood
Miniplanets – magpi.cc/miniplanets
Devwill Too – magpi.cc/devwilltoo
Arkagis Revolution – magpi.cc/arkagis
L'Abbaye des Morts – magpi.cc/labbaye
Old Towers – magpi.cc/oldtowers
Irena: Genesis Metal Fury (demo) – magpi.cc/irena
Cave Story MD – magpi.cc/cavestory

Miniplanets
3D platformer for an old console
Sik
Platformer
Add a blurb Remove...

Arkagis Revolution $14.99
A top-down shooter for Sega Genesis/Mega Dri...
Arkagis
Shooter
Add a blurb Remove...

NESertBus
NES demake of Penn and Teller's DesertBus
Redslash12
Simulation
Add a blurb Remove...

L'ABBAYE DES MORTS (MEGADRIVE/GENESIS) (POR 002)
L'Abbaye des morts Megadrive/Genesis by Mun
Playonretro
Platformer
Add a blurb Remove...

NEW Papi Commando Tennis Demo BEX - Megadrive
Papi Commando Tennis Demo "New" - MD
Studio Vetea
Sports
Add a blurb Remove...

Escape 2042 - The Truth Defenders $12
2D platformer \Atari STe / Gameboy / PC / Sega ...
Orion_
Action
Add a blurb Remove...

Papi Commando Remix Free Version - Megadrive
Studio Vetea
Action
Add a blurb Remove...

Break An Egg (0h Game Jam)
A Genesis / Megadrive game made in 60 minute...
Dr. Ludos
Action
Add a blurb Remove...

30 Years of Nintendon't
Dr. Ludos
Add a blurb Remove...

Break An Egg
A new arcade game for Megadrive / Genesis
Dr. Ludos
Action
Add a blurb Remove...

Tank You
Inspired by "Battle City" (Namco) - For a Game Ja...
tetsuro
Shooter

AA6061
fedorgames
Platformer

manganga
puzzle platformer game for windows and mega d...
nemezes
Adventure

MegaXmas'89
MegaXmas'89 is a new Christmas themed skiing ...
Nekojita
Sports

Perlin and Pinpin - The Robin Tower (Home)
SEGA Megadrive / Genesis game - Perlin & Pinpin
LIZARDRIVE
Adventure

▷ New developers publish games for classic consoles on popular indie platform itch.io

Top Tip 👍

Handheld paradise

If you'd rather build a handheld console, then that's a very viable prospect using a chassis such as the Retroflag GPi Case or the Waveshare Game HAT.

01 Image your RetroPie drive

Download the Raspberry Pi Imager for your operating system from **magpi.cc/imager**. Insert a microSD card – 8GB will be fine if you plan to limit yourself to 8- and 16-bit games, but if you want to emulate more powerful consoles in future, a 32GB card is a good investment.

Run Raspberry Pi Imager and pick RetroPie from the choose operating system list. You want the most powerful Raspberry Pi you can lay your hands on, but a Raspberry Pi Zero will do the trick if you stick to emulating relatively early systems, and is great for embedded console projects.

Choose your microSD card, click Write, and give the software permission to overwrite any data on the microSD card. Wait for the image to be downloaded and flashed.

Itch.io console games

Use these links to find new games for classic consoles on itch.io:

Mega Drive games: magpi.cc/itchmd
NES games: magpi.cc/itchnes
Master System games: magpi.cc/itchms
SNES games: magpi.cc/itchsnes

02 Plug it in, baby

Insert the microSD card and connect Raspberry Pi to a keyboard, mouse, and monitor. If you've got controllers or joysticks, plug them in before you power up.

After the image boots, you'll be prompted to assign your gamepad's buttons, if you have one. Trigger buttons on some controllers – notably Xbox 360 compatible gamepads – may not register when pressed. Press and hold any other button to skip configuring them for now. If you make a mistake, you'll be able to go back and correct it when you get to the end of the configuration list.

03 Fix your triggers (optional)

If the triggers are unresponsive on your Xbox 360 compatible controller, you should update the xpad driver. Go to RetroPie configuration and select RetroPie Setup. From the ncurses menu, select Manage Packages > Manage Driver Packages > 847 Xpad Driver, then Update.

Exit back to the main EmulationStation interface and open the Menu. You may find that this has been remapped from Start to the Right Trigger button after the update. Scroll down and select Configure Input.

CONFIGURING

GAMEPAD 1

HOLD ANY BUTTON TO SKIP

⊙ RIGHT ANALOG UP	AXIS 4-
⊙ RIGHT ANALOG DOWN	AXIS 4+
⊙ RIGHT ANALOG LEFT	AXIS 3-
⊙ RIGHT ANALOG RIGHT	AXIS 3+
🎮 HOTKEY ENABLE	BUTTON 10

OK

It takes a little getting used to, but EmulationStation's controller configuration tool means that RetroPie can handle almost any gamepad you want to use with it

◼ Set me up

04 With your controller configured, you'll be taken to the main interface. You won't see any emulators on offer until you've copied over games for them to play. Press A on RetroPie to enter the config menu.

Select WiFi. Press OK at the following menu to go on to connect to a wireless network. Choose from the network list and enter the network key. Select Exit to return to the main EmulationStation config menu.

Some 1920×1080 displays will show a black border. If this is the case, select raspi-config. Go to Advanced Options, then Overscan – this will get rid of the black border. Select No to disable overscan compensation. You'll need to reboot for this to take effect.

◼ Get some ROMs

05 Before we go any further, you'll need some games to run on RetroPie's suite of emulators. For our first NES ROM, we'll grab the Oliver Twins' Mystery World Dizzy. Go to **yolkfolk.com/mwd** and click Download. To test Mega Drive emulation, go to **arkagis.com** and click 'Download trial

> ❝ It's easiest to download ROMs on another computer and copy them across ❞

version' to take Arkagis Revolution's great rotating field navigation for a spin.

It's easiest to download ROMs on another computer and copy them across your local network to RetroPie's Samba share at **retropie.local** using your file manager. Each console has its own subdirectory under the **roms** directory. Windows users should ensure that network discovery is enabled.

Four modern NES games

These new NES games are excellent examples of modern retro game development:

Micro Mages – magpi.cc/micromages
From Below – magpi.cc/frombelow
Wolfling – magpi.cc/wolfling
Legends of Owlia – magpi.cc/owlia

◀ Micro Mages is a commercial modern NES game with fantastic graphics and tight single- and multiplayer gameplay for up to four people

Top Tip 👍

What's a ROM?

ROM (read-only memory) files are data images of a non-rewritable storage medium, usually a game cartridge or – more rarely – computer or console firmware.

Direct download ROMs

Although it's easiest to copy ROMs over from another computer, you can just download them at the command line of your RetroPie box if you have the URL. Press and hold **F4** to exit to the command terminal. You can download the ROM files directly to their directories using wget:

```
wget -P /home/pi/RetroPie/roms/nes/ http://yolkfolk.com/
flash/mwdidd.nes
```

Restart EmulationStation by typing exit at the command prompt. If you'd rather just download all your files to a single location and move them later, the Midnight Commander file manager accessible from the Configuration menu makes this fairly hassle-free too.

06 **Time to play**

Back on Raspberry Pi, restart EmulationStation: press Start on your controller, select Quit, then Restart System. Restart the interface every time you add games to force it to re-check its ROM directories.

If you have a keyboard connected, it's quicker to press and hold **F4** to quit to the command line, then type **exit** to restart EmulationStation.

As you scroll to the left or right, you should see logos for the NES and Mega Drive. Press A to enter the menu, then press A while highlighting the game you want to play. Right and left directional controls switch between different consoles.

▲ Old Towers is a new homebrew Mega Drive game, available as a digital download or even as a cartridge!

07 **Shortcuts, mods, and fixes**

Remember the Hotkey you defined during controller configuration? You'll be using that a great deal, as it serves as a mode switch for controller shortcuts. You'll find more info at **magpi.cc/hotkeys**, but these are the most useful:

Hotkey + Start – quit the game

Hotkey + Right Shoulder – Save

Hotkey + Left Shoulder – Load

Hotkey + B – Reset

Hotkey + X – Open quick menu for save states, screenshots, recording and similar

If you don't get any audio from Raspberry Pi 4, make sure the HDMI lead connecting your monitor is plugged into the HDMI 0 port, nearest to the power connector. ▥

YOUR ENEMIES AWAIT

Bring home all your favorite gaming villians and conquer the universe on your own full-size arcade console from the comfort of your private galaxy.

ICONIC ARCADE®

Cloud Gaming with
Raspberry Pi 5

Play the latest AAA console games with only a Raspberry Pi 5 and a good internet connection. No downloads, no updates. Here's how

PJ Evans

MAKER

PJ is a writer, software engineer and tinkerer. He really enjoyed the testing phase of this article.

mastodon. social/@mrpjevans

Did you know your Raspberry Pi can also double as an Xbox or a high-end gaming PC? You can now play the latest games in full HD using just your humble credit-card-sized computer. Your inputs (controller, keyboard) are sent to the remote 'rig' and a video stream of the game is sent back. Recent improvements mean lag is barely perceivable and the video quality is excellent. It's not quite as easy as powering up and jumping into *Baldur's Gate III*, however, and you'll need to make a few decisions along the way. This tutorial will walk you through it all.

01 Choose your platform

There are a handful of cloud gaming services available. Out of all of them we've chosen Xbox Game Pass Ultimate as a great service for our Raspberry Pi 5. For a monthly subscription, you can access hundreds of games that are instantly playable, and many top games are

included. We'll also cover other options such as Nvidia GeForce Now. Sadly, Sony's PlayStation equivalent is PC-only, and we encountered issues using Amazon's Luna. Xbox and Nvidia can work with Raspberry Pi as they both support streaming to the browser, including Chromium.

02 Choose your hardware

A good controller is essential for any gaming, cloud or otherwise. We've chosen the 8BitDo Pro 2. Not only is it a great Bluetooth controller, but it's also proven with Raspberry Pi hardware and approved by Xbox. You can also use an official Xbox controller if you prefer. You'll also need to interact with a keyboard and mouse. Of course, you can use a regular wired keyboard and mouse connected to the USB ports, but if you're setting up a gaming environment you may want to consider a wireless keyboard such as the Logitech K400 series which combines a keyboard and trackpad in one.

03 Initial setup

We're going to need the 'full' version of Raspberry Pi OS, complete with desktop, as we need the Chromium web browser. Using the Raspberry Pi Imager (**magpi.cc/imager**) select Raspberry Pi OS (64-bit) and write the image to your SD card to get the latest evolution of the operating system. We strongly recommend using a wired internet connection to your Raspberry Pi if possible, for best results, but if you're using Wi-Fi, set it up now (and SSH too if you need remote access). Once booted and ready to go, check and install any software updates (or run `sudo apt update && sudo apt upgrade` in the Terminal).

You'll Need

> Fast internet connection (ideally fibre)

> 8BitDo Pro 2 controller **magpi.cc/pro2**

> Wireless keyboard/mouse **magpi.cc/k400**

▲ The 8BitDo Pro 2 controller is a perfect choice for this project, and is endorsed by Microsoft for Xbox Games Pass

You'll need to use the keyboard and mouse as well as a game controller, so choose a wireless option like these

04 Set up your controllers

The keyboard (with built-in trackpad) we have chosen is very simple to set up because it uses a dedicated USB dongle. Just plug in the receiver, and make sure the keyboard has some batteries and is switched on. Both it and the trackpad should work immediately. Bluetooth devices, such as our controller, need to be paired first. Click on the Bluetooth icon at the top-right of the screen and select 'Add Device'. When the scanning window appears, put your controller into pairing mode (on the Pro 2, press Start + X then press the sync button for a few seconds). The device should appear in the window. Now click 'Pair' to connect. You only have to do this once.

> ❝ Recent improvements mean lag is barely perceivable and the video quality is excellent ❞

05 Get an account

It is a necessity of cloud gaming that you have an account with the provider. In Xbox's case, this is Microsoft. So if you haven't got one, now is a good time to get yourself set up with a Microsoft ID. Go to xbox.com, click 'Sign in' then 'Create account'. Go through the steps to get set up, and don't forget to choose a strong password and multi-factor authentication. Now you can log into xbox.com and get ready to start gaming.

Thanks to HDMI, you can do your gaming on the big screen

With cloud gaming our favourite diminutive computer can hold its own against bigger machines

06 Sign up for Game Pass Ultimate

Xbox's Game Pass service comes with different tiers of access. Cloud gaming is only provided with the Ultimate package, which at time of writing is £12.99 per month with a £1 offer for the first month. For your money you get full access to over 300 games, from classics like *Fortnite* and *GTA*, indie greats like *A Short Hike* and newer top titles such as *Starfield* and *Football Manager 2024*. Sign up will require credit card details which will be debited monthly. You can cancel your subscription at any time. Get yourself signed up and then you're ready to start gaming.

07 Try it out

Open up Chromium on your Desktop and go to xbox.com. If you haven't already, log in now. When you are returned to the front page, click 'Games' then 'Cloud games'. You'll have a selection of games chosen for you, or you can scroll to the bottom and click 'See all games'. Choose a game and then click 'Play'. You may get a warning that a controller has not been detected.

Top Tip

Choose the right mode

Using the 8BitDo Pro 2? Always begin by pressing the X + Start buttons to ensure maximum Xbox compatibility.

You'll need to sign up for Xbox Games Pass Ultimate. It comes with access to over 300 games

If so, click A on your controller and it should clear. If that doesn't work, check your pairing. You'll now be put in a queue for the first available rig.

08 Playtime!

Your wait time will depend on the game's requirements and popularity. In our testing, most games kept you waiting no more than a couple of minutes, but popular and resource-hungry games can see waits of 30 minutes or more at peak times. If this happens, try choosing another game. The wait screen will keep you informed of progress. When the game starts, it will be as if you were in front of a real console and you play accordingly. Pressing the 8BitDo controller's 'heart' button will bring up an overlay so you can end the game or adjust settings.

Our experience streaming games with Luna wasn't good, but it might work for you

09 Troubleshooting

Now you will find out whether your setup has a good enough connection. The service will try to adjust video quality, prioritising speed over resolution. If you find it inadequate, try moving from Wi-Fi to a wired connection. Check whether your sibling is streaming HD video while uploading 5,000 selfies. If nothing works, contact your internet provider to see if there are any faster options available to you. We've found that fibre-based services are the only way to get close to a real console experience. If you're finding response times laggy, nearby radio interference may be the cause, and using an Ethernet cable to connect to your router can fix this too.

10 Boot straight in

To finesse your Xbox experience, we can configure Raspberry Pi OS so you can boot straight into a full-screen browser. Open up a Terminal window or SSH into your tiny gaming rig. Now issue this command:

```
nano ~/.config/wayfire.ini
```

Go to the bottom of the file and add this as a single line:

```
[autostart]
chromium = chromium-browser https://www.
xbox.com/en-GB/play?xr=shellnav/ --kiosk
```

▲ Nvidia's GeForce Now platform comes with some great demos for free, such as *Lego Bricktales*

```
--noerrdialogs --disable-infobars --no-first-
run --ozone-platform=wayland
```

Save the file and quit Nano (**CTRL**+**X**). Reboot, and Chromium will start in fullscreen mode automatically, and go straight to the Xbox home page.

11 Nvidia GeForce Now

If you want to try cloud gaming completely for free, check out Nvidia GeForce Now, a similar service to Xbox Game Pass. It has a free tier that allows you to access a small subset of games, including an excellent demo of *Lego Bricktales*. It runs in the browser just like Xbox's service, but can link in with popular game libraries such as Epic and Steam to allow access to your purchased games (if supported). You can also choose a paid tier to get access to the games libraries and better hardware. Visit **magpi.cc/geforcenow** to check it out.

> The service will try to adjust video quality, prioritising speed over resolution

12 Other platforms

We also tried Amazon Luna, which is a similar offering to Nvidia's. However, although streaming worked well, we found it struggled with controller input, with seemingly random keypresses ruining the gaming experience. We were unable to establish whether this was a temporary glitch, however, so do try Luna if you have Prime membership. If you're looking at a permanent gaming setup, don't forget that cloud-based streaming is not your only option. Raspberry Pi is a brilliant platform for retro gaming (check out RetroPie) or even having a go at making your own with Python PyGame. Whichever way you go, have fun!

Top Tip

Steam Link

If you want to play your PC games in another room, check out Steam Link for Raspberry Pi.

RETRO GAMING HARDWARE

REVIEWS OF THE TOP KIT FOR RETRO GAMERS

8BitDo Pro 2 **Controller**

▶ 8BitDo ▶ **magpi.cc/pro2** ▶ £40 / $50

Professional-grade video game controller that works a treat with Raspberry Pi. **Lucy Hattersley** flexes her thumbs

SPECS

POWER:
1000 mAh lithium-ion battery or 2 × AA batteries

CONNECTIVITY:
USB-C, Bluetooth 4.0

DIMENSIONS:
153.6 × 100.6 × 64.5 mm, 228 g

▼ The Pro 2 controller features all the buttons from a modern console gamepad

8BitDo is a company that's been making a name for itself in the retro gaming sphere, supplying quality game controllers and conversion kits at a good price.

The firm recently sent us a box of interesting things to look at, and we decided to start here, with the Pro 2 controller.

Reminiscent of a PlayStation DualShock, the Pro 2 controller has two analogue sticks, a D-pad, four buttons, four triggers, two Pro-level back

buttons, Select, Start, Star, Heart, and Profile button. It's certainly not short of a button or two.

If all that wasn't enough, there is an 'SADX' Mode switch underneath that swaps between four different modes: Switch, Apple, Android, and Windows.

It comes with a long USB-C cable and 1000 mAh lithium-ion battery with "20 hours of battery life."

In terms of value, £40 is not particularly low-cost in the world of Raspberry Pi, but it is good value when stacked up against its immediate rivals: a Sony DualShock will cost you £50 and an Xbox One Controller starts at £55 (without a rechargeable battery). So, this is cheaper than either. But is it better?

Setup was a breeze. We used the Windows (X) setting on the back and started with a direct USB-C connection. We then held down the Pair button and synced it up with 'Add Device' in the Bluetooth settings in Raspberry Pi OS.

Support in Raspberry Pi OS is game-dependent, although we had a blast in Super Tux Kart and Doom.

We moved onto classic games with Batocera.linux (**batocera.org**) which is a new retro gaming distribution that we'll be talking more about in future. Setup was even easier there, requiring us only to plug in via the USB-C and hold the Pair button. RetroPie was equally easy to set up, mapping the buttons on the controller during the setup process.

For a more modern experience, we tested it out with Xbox Cloud gaming (**xbox.com/play**). This enabled us to use all the analogue sticks and

◀ The SADX Mode Switch enables you to adjust the controller mapping to fit a range of consoles, devices, and computers

❝ It integrates neatly with Raspberry Pi ❞

triggers with some of the latest 3D masterpieces. Again, we had no problems.

Button combinations can be mapped to the two Pro buttons on the rear; sadly, the software to control them is only available for Windows, macOS, Android, and iOS. It's a shame you can't do the setup via a web or Linux app.

There are a few quirks. You switch off the controller by holding down the Start button for three seconds, and you might find using Windows (X) not immediately obvious over the other settings. But really there's nothing here that a read of the supplied instruction manual won't clear up.

Holding its own

Build quality of the Pro 2 controller is superb. It's easy to grip and buttons have a nice responsive click with no sponginess. The analogue sticks are weighted well and spring cleanly back to the centre. It's certainly a step above the usual fare for a third-party controller.

The Pro 2 integrates neatly with Raspberry Pi, and the Mode Switch means you can quickly transfer it to any other consoles or computers that you might be using.

We really have no hesitation in recommending this one. ❚

▶ Two Pro buttons, found underneath the gamepad, can be mapped to button combos using separate (non-Linux) software

Verdict

A fantastic controller for a good price that works across a range of Raspberry Pi games, apps, and distributions. Easy to set up and use. Shame the Ultimate Software isn't available in Linux, though.

9/10

8BitDo **Arcade Stick**

SPECS

DIMENSIONS:
303 × 203 × 111.5 mm

WEIGHT:
2.1 kg

CONNECTIVITY:
Bluetooth, RF (through included dongle), wired via USB

MACRO BUTTONS:
2

BATTERY:
1000 mAh Li-ion, 30–40 hour play time

▶ 8BitDo ▶ **8bitdo.com** ▶ £78 / $90

A classic console colour scheme on a very modern arcade stick. Can **Rob Zwetsloot** put it through its paces? Sure he can!

The market for high-quality arcade/fight sticks has been an interesting one over the last decade – while in the West arcades are few and far between, there are people still interested in playing games like they did in the arcade. Whether they're hobbyists enamoured with the classic style, or hardcore players in the fighting game community (FGC), there will always be people looking for one. Raspberry Pi retro gamers are no different.

8BitDo have answered with their own stick simply called Arcade Stick. Like their other controllers, it's fairly platform-agnostic, making it playable on a Switch or PC – and that includes Raspberry Pi. It uses XInput, the Xbox 360 controller API, allowing you to use it with very little hassle in a vast majority of games, whether you're using RetroPie or Steam Link. The eight face buttons map the major buttons on all modern controllers, and it includes two programmable macro buttons, switches to change how the stick is defined, and a couple of extra function buttons for Start, Select, Home, etc.

Connect it your way

There are three ways to connect the stick to a Raspberry Pi, each of them good in its own way. While technically the fiddliest way to connect is

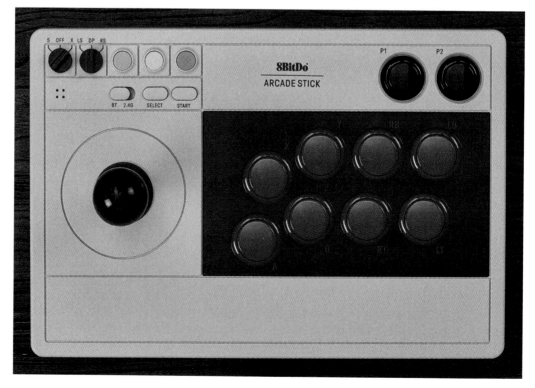

▶ The buttons use a tried and tested layout, while the switches are unique to the way the stick works

via Bluetooth, it does mean you get a wireless connection without having to use a dongle. Speaking of dongles, there's a little RF receiver tucked behind a door on the rear of the stick that come pre-paired. Plugging it into a Raspberry Pi

> ❝ It feels weighty and sturdy, and doesn't slip easily if you plan to place it on a table instead of your lap ❞

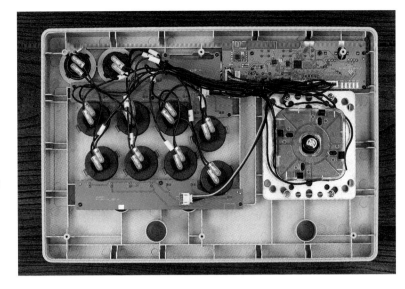

▲ Like all good modern fight or arcade sticks, you can easily swap out and mod the buttons and stick

and flicking over to 2.4G allows you to immediately use it wirelessly without any Bluetooth hassle, and it works really well. It also comes a USB-C cable which can be stored before the same door, which also houses the wires connection – just in case you need to perform that frame-perfect parry of Chun-Li's Houyoku Sen without wireless lag.

It feels weighty and sturdy, and doesn't slip easily if you plan to place it on a table instead of your lap. Like a lot of modern arcade sticks, you can also customise and swap out the buttons and joystick. The included square-gated joystick and buttons are fine, but feel a little loose for our liking, and if we were going to start using it in a professional setting, we'd definitely replace the

internals with some high-end Sanwa parts for more satisfying presses.

For the price though, you're getting a very flexible, and still very good, arcade stick. Sadly, like the Pro 2 controller we reviewed last issue, the macro buttons cannot be programmed on Raspberry Pi. However, there is a turbo button that you can use at any point, so there is some minor customisation you can do.

Overall, we love it. It feels like a quality piece of kit, and performs as such too. We're also keen on the use of switches rather than cheat code-esque button combos to change functionality, like in the last generation of 8BitDo products. Now, excuse us as we go take down Shadaloo. 🎮

◄ The included RF dongle makes it very easy to plug and play remotely

Verdict

An exceptionally solid arcade stick let down by some mildly fuzzy buttons, which will admittedly only bother a minority of players.

9 /10

PiBoy **XRS**

SPECS

INPUT:
2 × analogue sticks, 1 × D-pad, 7 × face buttons, 4 × shoulder buttons

PORTS:
3.5 mm audio jack, mini HDMI out, 2 × USB 2.0, 2 × USB 3.0, 1 × Gigabit Ethernet

DIMENSIONS:
162 × 93 × 36 mm

DISPLAY:
3.5" IPS LCD

▶ Experimental Pi ▶ **magpi.cc/piboyxrs** ▶ £125 / $150

A more ergonomic handheld experience, the PiBoy XRS takes the best parts of a lot of classic consoles. **Rob Zwetsloot** has a play

We were big fans of the PiBoy DMG, a Raspberry Pi 4-compatible handheld in the shape of the classic Game Boy, albeit with a few more buttons and an analogue stick. It's a great form factor for nostalgia purposes. However, even Nintendo switched to a horizontal format eventually. This brings us to the PiBoy XRS which, while quite like the DMG version, is definitely a superior product.

Taking design notes from the original Game Boy, Neo Geo Pocket, and just about every portable game system since the release of the Game Boy Advance, this updated kit also whacks on a second analogue stick for much improved playability. Instead of two shoulder buttons, there are now four, and while that does mean two fewer face buttons, for the vast majority of games it's a layout we prefer. It turns out there's a reason why handhelds went horizontal after the Game Boy.

Hook it up

Installing a Raspberry Pi 4 is very simple – with the supplied screwdriver, you just need to take the eight screws off the back and then slot Raspberry Pi into the GPIO pins. There's some thermal paste for the processor, and an optional adapter for the mini HDMI out. Four of the screws on the back-plate go through the mount holes on Raspberry Pi, making for a very elegant reassembly process.

Power and headphone jacks are routed to the underside of the console, with rechargeable batteries, adding up to 5600 mAh, connected to that USB-C power port for hours of play time. The standard USB and Ethernet ports are then left open at the top of the XRS for easy, if not perhaps slightly awkward, access for peripherals and memory sticks. Those pop out of a fake cartridge slot, which is a nice little design touch, especially paired with the hand-detachable cover that allows you to change out microSD cards without having to get the screwdriver back out.

Playing with power

Like DMG, the whole package is solid and feels good to hold – we find some handheld Raspberry Pi consoles can feel a little flimsy and, while this may be a little heavier than those, we appreciate the added heft.

▲ There's easy access to the microSD port, and plenty of ventilation for Raspberry Pi 4 too

▲ The joysticks remind us of the smaller ones on the Switch Joy-Cons

Playing games is incredibly simple. Once the PiBoy image is installed, you can game basically straight away and as it's the familiar RetroPie interface, you can easily customise it to your liking. Compatible games run smoothly thanks to the extra oomph of Raspberry Pi 4, and everything on the screen looked lovely as we played. Our kit came with the mini HDMI adapter, and we had

◄ Access to the USB ports is handy however you play it. There are much fewer use cases for the Ethernet port, though

> ❝ Once the PiBoy image is installed, you can game basically straight away ❞

no problems playing games on a bigger TV with it plugged in either – although, due to the graphical prowess of some games, it did look better on the smaller screen with a higher pixel density.

Also, as it has RetroPie, it means you can install Steam Link, so it's a pretty cool alternative to a Steam Deck. ⧉

Verdict

The perfect retro handheld system is finally here, and you can even use it to comfortably play modern games as well.

10/10

Alpakka

▶ Input Labs ▶ **magpi.cc/alpakka** ▶ Free + cost of components

SPECS

LICENSES:
Open-source firmware, Creative Commons hardware

INPUTS:
2 × gyro sensors, 13 × face buttons, 2 × sticks, 1 × scroll wheel, 6 × shoulder/ grip buttons

OTHER FEATURES:
Customisable, Raspberry Pi Pico-powered

▲ It's powered by a Pico inside which allows the whole controller to work like a standard XInput controller

A DIY controller that can be used like a mouse or a gamepad with extra buttons. **Rob Zwetsloot** get to grips with it

DIY controllers are becoming more and more popular in the gaming space, especially when it comes to making accessible controllers for folks with disabilities, and 3D printing has really made it easy to roll your own input system.

Alpakka has taken all this and gone a little step further, creating a customisable controller base for (mostly) free and showing you how to put it together. We say mostly free because you do have to buy some extra components and get the PCB printed. However, even with a handful of components, you're not spending more than a tenner, especially if you have a Raspberry Pi Pico lying around to power it.

Ours came pre-assembled but the instructions are very clear, requiring you to flex only a few soldering skills to put it together.

Custom controls

Once everything is soldered, construction is fairly straightforward. Parts slot in and are tightened in place with a few screws, with the hardest part

being the final sandwiching of the parts to finish the controller.

Build quality is quite dependant on how you print the controller, and also if you decide to customise it before sending the STL over to your printer. With all the electronic parts inside, it does have a satisfying weight and, thanks to the compact design, it doesn't feel loose and creaky like other 3D-printed handhelds and controllers we've used in the past.

Game on

As for how it plays – if you've ever used 3D-printed controls before, you'll know that they don't afford as much comfort as other controllers with softer and smoother plastic. The right stick isn't really supposed to act like a traditional right stick, so it sits there as an awkward cube that doesn't give the best look in first- or third-person games for aiming.

The left stick also feels a little clunky, and the triggers are hard to use as, thanks to the extra leverage you have on them, it's difficult to feel the click on the button. They're also not analogue like on modern game consoles.

One of the reasons the controller is like this, though, is that you can also use it like a mouse thanks to advanced gyro functions – in fact, the scroll wheel on the controller is supposed to mimic one on a mouse. We found it a little awkward to use as we're not very used to it, but it definitely

> ❝ With all the electronic parts inside, it does have a satisfying weight ❞

works better than older systems we've tried. It can also easily be turned off and on again, which is a bonus.

For the price of putting this together at its most basic, it is very good though. It's great for more retro games, too, that don't require analogue controls and, due to its DIY nature, you could easily find more comfortable replacement buttons if you wanted to head down that route. 🅼

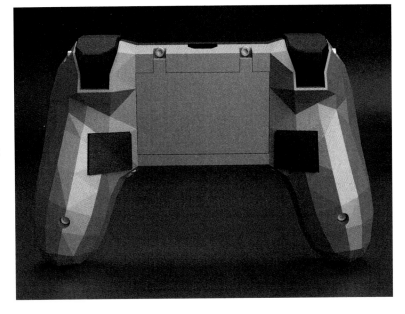

▲ Rear grip buttons are also included, just in case you need more inputs

◀ It's a very cool-looking controller with even cooler features packed inside

Verdict

We really like the idea of Alpakka, but if you want to use it as a main controller, you will need to start customising it yourself.

7 /10

Kitronik ZIP96
Retro Gamer

▶ The Pi Hut ▶ **magpi.cc/zip96** ▶ From £36 / $38

Learn to code with this Pico-powered handheld. By **Phil King**

Retro gaming is fun, so how about creating your own old-school games? The ZIP96 offers the opportunity to learn to code and enjoy playing the results on a basic handheld console. It doesn't even have a case, although you could always design and 3D-print one.

The bare-bones PCB features four standard tactile push-buttons for a directional pad on the left, with a couple of action buttons on the right. Between them is not a screen, but a 12×8 LED matrix with addressable RGB pixels. Below this are twin female headers to plug in your Raspberry Pi Pico – one isn't supplied. Between the headers is a tiny piezo buzzer for rudimentary sound (beeps with settable pitch).

On the rear of the board you'll find a vibration motor that spins at high speed to create the kind of rumble effect you get on some commercial games controllers. You'll also find three AA battery holders for portable power – more on that later.

Pico programming
Too basic to stand up as a pure gaming device, the ZIP96 is aimed more at the educational market. For this purpose, there's a series of ten lessons available as downloadable PDFs, taking learners through the fundamentals of object-oriented programming in MicroPython as they create a simple maze game, called the A-Mazing Game. The well-structured, step-by-step lessons cover aspects such as classes, objects, functions, loops, and threads. It's a very good introduction.

As usual, you connect Pico to a computer via USB and use the Thonny IDE to start coding. You'll need to install the ZIP96 MicroPython library, which is available in Thonny's PYPI package manager – just search for it there. This makes it relatively easy to read and trigger the ZIP96's various components by name. The pixels on the LED matrix can be addressed using X and Y co-ordinates or by their ID number, from 0 to 95 (numbered left to right, top to bottom).

In the ZIP96 GitHub repo (**magpi.cc/zip96git**), you'll find files for the A-Mazing Game at various stages of construction, to go with the lessons. There's also a classic Snake game to try out, along

SPECS

FEATURES:
12×8 RGB ZIP LED matrix display, 6 × input buttons, piezo buzzer, vibration motor, on/off switch, 3 × breakout headers

POWER:
3 × AA battery holders

DIMENSIONS:
141 × 60 × 24.6 mm

with another code example written in C: a platform game called Run Along, Jump & Jump. There's even a wireless two-player version of the latter for two ZIP96 boards equipped with Pico Ws.

Powering the matrix

One curious feature we came across while coding is that, for some reason, USB power can't be used to run the LED matrix. You need to have AA batteries inserted and the device's slider switch set to on. Not a major issue, although we did mistakenly think we might have a dud unit for a little while. The other downside of this setup is that you can't use just a USB power bank for portable play, which is a shame.

> ❝ Too basic to stand up as a pure gaming device, the ZIP96 is aimed more at the educational market ❞

Another issue we encountered was that you need fully charged batteries for the LED matrix to light up in the right colours. With low batteries, we found that the blue element didn't show up, so cyan pixels looked green, for instance. ▣

▲ On the rear of the board are three AA battery holders, a vibration motor, and another breakout header

▲ Playing a simple game on the ZIP96 is quite fun, but it's mainly about learning to code

Verdict

An interesting and fun educational tool. The simple LED matrix display limits the possibilities for creativity, but does make it easier learn the fundamentals of coding.

7 /10

◀ The ZIP96 doesn't come with a Pico. Two breakout headers at the top enable you to add optional shoulder buttons

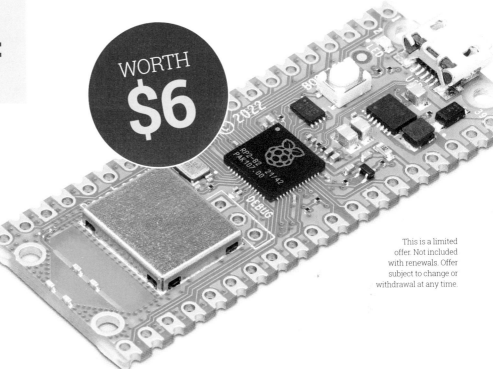

RETRO PROJECTS

INSPIRING PROJECTS AND STEP-BY-STEP BUILDS

Fancy Octopus Arcades

Inspiring the next generation of gamers led to a business creating custom arcade housings, discovers **Rosie Hattersley**

MAKER

Shonee Strother

Shonee swapped a role as creative director and fulfilled a childhood dream to build gaming arcades, with assistance from his "highly talented" six-year-old son.

magpi.cc/ fancyoctopus

"**W**e're in the golden age of emulation,**"** declares **Shonee Strother.** He doesn't have a standard technical background, but he's been a gamer all his life and, when the pandemic put his role as a creative director on hold, he began looking for new hobbies. Shonee and his six-year-old son Wolf started their Raspberry Pi adventures "by modding an arcade 1up with a Raspberry Pi 3B." Wolf is extremely interested in gaming, like his dad, who felt that his son's young age meant "there's so much gaming history he's missed. Giving him a slice of that was super important." When Shonee documented their work on Instagram, people began to take notice. Their bespoke gaming arcade business, Fancy Octopus Arcades, "just sort of took off from there."

Bijou is beautiful

Brooklyn resident Shonee started following a few Reddit and Facebook pages where he "saw folks making huge gaming rooms, loading them up with

▶ The unique retro arcade builds have proved extremely popular, with handheld designs planned soon

arcade cabinets, and realised that for some of us in smaller apartments, that wasn't a possibility." Shonee was also frustrated that people were capitalising on the scarcity of arcade cabinets, "snagging all the 1up machines they could find and price gouging them to death." He decided to address pent-up demand for games arcades, focusing on designs that would fit a New York-sized apartment.

The idea was to make small, mini cabinets, packed to the gills with games, easily plugged via standard HDMI for video audio output. Shonee also wanted them to be highly personalised art pieces "that a person could be proud to put on a display shelf or coffee table." Raspberry Pi's small profile and extremely robust performance helped him "pack a lot of oomph into a very small package," he confirms. He also loves being able to easily flash images directly onto the SD card with little or no manual coding changes. As a result, he can focus on each cabinet's artistry.

Daring designs

Although homemade retro games cabinets are a popular build, Shonee's stand out because he designs and creates all the parts himself. The 3D printing, silkscreen, vinyl wrap, and decoupage work is all done in-house, depending on the build. Shonee works closely with clients, creating a design around their vision. "No two builds are ever alike, and no two designs are ever repeated," he says of what is now a two-year-old business. "I've had so many folks ask me 'can you build me a copy of that Donkey Kong deck?' My bank account absolutely hates that I have to say no." The unique elements that bring his designs to life include gear sticks masquerading as swords for his Samurai

Although each build is different visually, inside there's a Raspberry Pi running RetroPie games emulators and ArcadePunks' front end software

The portable arcade cabinets use zero delay encoders wired to 28 mm LED arcade buttons and also feature some powerful internal speakers

The custom-designed exteriors involve skills such as vinyl cutting, painted artwork, 3D-printed elements, and unexpected extra features

Quick **FACTS**

> Each arcade Shonee builds has a secret, unannounced feature

> The Zelda one has a rainbow LED rupee-filled chest on top

> It's a music box that plays the Zelda fairy theme

> The arcade can be updated remotely and run off rechargeable batteries...

> ...Ideal for participating in a recent MVC2 battle in Central Park

▲ Arcade cabinet designs feature elements from the client's favourite games

▲ Following the success of the bespoke arcades, Shonee is keen to branch out into custom handheld consoles

◀ This RetroPie arcade features a barrel-shaped planter from a garden store

Pimp your arcade

Build the arcade cabinet, either by downloading and printing designs from a site like Thingiverse or creating a custom wood build. You'll need a Raspberry Pi, zero delay encoders for buttons and joysticks, speakers, a mini amp, and some creativity.

01 Wire up the unit

Connect zero delay encoders to 28 mm LED arcade buttons. For internal audio, add a mini amp and stereo speakers mounted to the control deck.

02 Allow for updates

Connect Raspberry Pi to an external Ethernet port for any updates or system changes. Install a USB hub to power the LEDs, and attach a mini keyboard for any edits that may be needed.

03 Testing, testing

Flash-update RetroPie and run it to make sure the front end works with the library you've installed. Depending on the display you're using, check for scan lines, and adjust frame rates and the aspect ratio.

Jack build, the handle from a Panasonic boom box found at a Chinatown flea market and used in his Basquiat/Keith Haring build, and a Donkey Kong barrel refashioned from a garden centre planter. "Basically, I use whatever I can." More often than not, he uses software from **Arcadepunks.com**. Their front ends are "stable, responsive, and rarely cause problems down the line," Shonee says.

"Some folks really want to focus on retro arcade gaming; some folks are more focused on having a game preservation library of console games from their past. There are always some adjustments that need to be made. Luckily, with the versatility of Raspberry Pi, it's pretty easy." Each Fancy Octopus Arcade typically takes about a month to complete, with costs varying depending on the brief.

"Without Raspberry Pi, this would just be a box with buttons on it," Shonee concludes. "The size, power, versatility, and ease of use have given me the ability to help the dream of making my hobby a career come to fruition." 🅼

▲ The handle for this Haring vs Basquiat build is from a 1980s Panasonic boom box that Shonee bought at a NY flea market

Sol-20
Terminal Computer

A fleeting glance at a gorgeous limited edition computer 45 years ago
resulted in a marvellous Raspberry Pi 4 build, **Rosie Hattersley** learns

MAKER

Michael Gardi

Michael is a retired
software developer
from Ontario,
Canada who enjoys
spending his free
time on retro
computer builds
and whatever
else the heck he
feels like.

mikesmakes.ca

Laying claim to be the first fully-assembled
microcomputer with a built-in keyboard and
television output, the Sol-20 was launched
back in 1976. "It had more in common with the
Altair 8800s and IMSAI 8080s of the day, than it
did with the Apple and Commodore computers that
were soon to follow, despite looking more like the
latter", says Michael Gardi who, a full 45 years
later, built his own version, powered by Raspberry
Pi, since a version made using only vintage
components would be prohibitively expensive.

Michael's home is full of Raspberry Pi computers
performing tasks from NAS to OctoPrint server

and RetroPie games arcade, so it seemed obvious
to use our favourite single board computer for his
Sol-20 project. He chose a Raspberry Pi 4 because
he needed "the horsepower it provides to run the
emulator at speeds comparable to the original Sol-
20". He also realised the GPIO ports and option
to use a variety of HATs – particularly the voltage
shifter module – made Raspberry Pi a great choice
for this build.

In the end, the project took around four months
to complete, and cost roughly US $550.

Computer love

Michael wanted his Sol-20 reproduction to be "as
authentic-looking as possible, with the beautiful
walnut sides for sure". He'd always admired the
original computer's striking steel blue sides too, but
chose to 3D-print the case for his version, pointing
out that also makes it easier and cheaper for anyone
else following in his retro build footsteps.

Raspberry Pi came into its own when it came
to the software written for Sol-20. Recreating
the motherboard to do this and populating it
with vintage parts would have been prohibitively
expensive, Michael explains. Raspberry Pi runs
code that emulates the 8080 CPU at the heart
of the Sol-20. "On top of this, it emulates the
original hardware providing a display via the

▶ Other Sol-20
emulators exist, but
Michael wanted to
create one specifically
to run on Raspberry Pi

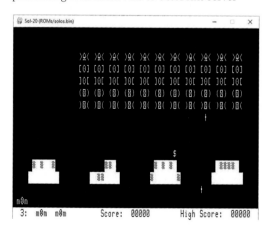

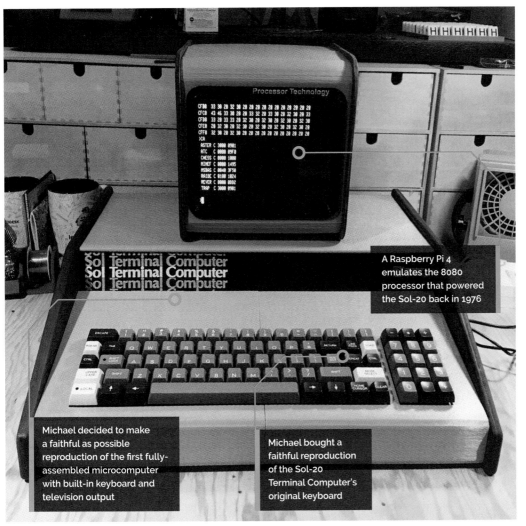

Processor Technology

A Raspberry Pi 4 emulates the 8080 processor that powered the Sol-20 back in 1976

Sol Terminal Computer
Sol Terminal Computer
Sol Terminal Computer

Michael decided to make a faithful as possible reproduction of the first fully-assembled microcomputer with built-in keyboard and television output

Michael bought a faithful reproduction of the Sol-20 Terminal Computer's original keyboard

Quick FACTS

> Only 12,000 Sol-20 Terminal Computers were produced

> Original models sell for $5000 or more

> Michael spotted one in a shop in Toronto in 1976

> And has hankered after owning one ever since

> His Sol-20 emulator runs on any Raspberry Pi

▼ Faithfully recreating the 1976 Sol-20 Terminal Computer involved learning to use a CNC machine in order to make the walnut sides

🔲 If you're thinking of your own reproduction of a classic, Michael warns you need to be prepared to be in it for the long haul 🔲

HDMI port and a virtual cassette interface for loading and saving programs." Michael used Python to write the Sol-20 hardware emulation software on top of an existing 8080 emulator that he found on GitHub (**magpi.cc/py8080**).

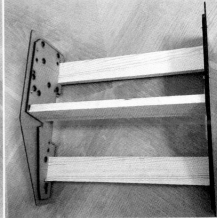

▲ Michael 3D-printed the reproduction Sol-20 case, but learned to use a CNC cutter to make walnut wood sides to match the original design

▶ Fellow retro computer builder Osiweb had designed an ASCII keyboard interface board for which Michael adapted code for Raspberry Pi and Sol 20

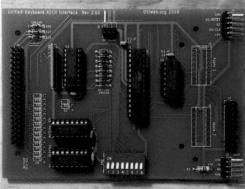

His Sol-20 emulator (**magpi.cc/ sol20reproduction**) can be run on Raspberry Pi 4 with a standard keyboard and display, but another Sol-20 fan had recreated a keytronic version of the original keyboard (**osiweb.org**), and Michael couldn't resist!

Screenwise, he bought a Pimoroni PIM372 8in XGA display. The full build details and hardware list can be found at **magpi.cc/sol20bom**.

New tricks

With detailed documentation about the computer he was trying to recreate available online, Michael says the Sol-20 project went smoothly. As a retro computer build veteran, he was able to apply lessons learned from previous projects such as the KENBAK-2/5 and DEC VT100 Terminal, both of which we've featured in *The MagPi*. However, he also challenged himself, learning how to use a CNC machine at his local maker space in order to recreate the Sol-20's "gorgeous" walnut side panels.

If you're thinking of your own reproduction of a classic, Michael warns you need to be prepared to be in it for the long haul. "Don't start unless you are passionate about the idea," he cautions,

before enthusing: "Break the project up into smaller pieces that can be finished in a relatively short time. Keep plugging away at it. Have fun!"

For further information on Sol–20 Terminal Computer, Michael recommends visiting **sol20.org**, where you'll also find Solace (SolAnachronistic Computer Emulation), a Windows version of Sol's interface. 🄼

▲ The reproduction Sol 20 keyboard looks great, but involves assigning key controls very carefully

◀ Although a facsimile keyboard exists, the keyboard layout is not an exact match

Recreate Sol-20

To build your own Sol-20 you'll need a Raspberry Pi 4, a 3D-printed or self-drafted computer case, an 8in 1024 × 768 × GA screen, 3.3 V to 5 V bi-directional voltage level adapter, USB to serial, and TTL adapters and, ideally, a keytronic keyboard.

01 Build or 3D-print the Sol 20 frame. Download the emulator software from **magpi.cc/sol20reproduction** and install it on Raspberry Pi 4 or another Python environment. Clone the repository and run python main.py.

02 Attach the 3.3 V to 5 V voltage adapter to Raspberry Pi 4, along with other hardware including power supply and fan.

03 Assemble the keyboard and keyboard emulator as per instructions at **magpi.cc/sol20bom**, then connect and test the keytronic keyboard and the LCD XGA screen.

ZX Spectrum
Raspberry Pi Cassette

Between jobs, one maker decided to push their Raspberry Pi skills and make a portable ZX Spectrum Raspberry Pi. **Rosie Hattersley** approves

MAKER

Stuart Brand

Stuart aka JamHamster is "an IT Root Cause Engineer for work and an avid (some would say obsessed) tinkerer in my off hours!"

@RealJamhamster

Stuart Brand was between jobs and decided to concentrate on pushing his skills by building Raspberry Pi projects: "I headed to the garage and embraced my inner nerd!" exclaims the maker of the ZX Spectrum Raspberry Pi Cassette. "I wouldn't have had a clue how to build any of this stuff before lockdown. It goes to prove that you never know what you're capable of until you give it a go."

Stuart's first computer, a ZX Spectrum, has a special place in his heart, so a Raspberry Pi project based around one seemed ideal. "They're still great machines!" he says of the beloved computer which celebrates its 40th birthday this April.

Stuart loves repairing and running real hardware as well as emulations and thought "it would be nifty to see if I could fit an entire ZX Spectrum emulator into a cassette tape shell." He now uses his ZX Spectrum Pi Cassette as a 'pick up and play' device whenever he fancies "a quick bash at some old school gaming."

Learn as you go

Prior to this project, Stuart had several retro makes under his belt and had made a tape emulator for an Arduino-based ZX Spectrum +2 that acts like a multi-cart tape. "Putting a whole Spectrum in a tape shell was the next logical step and an interesting challenge," he says. Being tight for space, he chose Raspberry Pi Zero W. He loves the fact both ZX Spectrum and Raspberry Pi's ARM processor were developed in Cambridge.

> ## Stuart wrapped Raspberry Pi in foil and went at it with a Dremel

Despite this, he describes himself as a haphazard tinkerer with little electronics experience, who plans everything in his head. "I don't have any schematics to share," he apologises, "and never measure anything." However, he makes paper mock-ups of everything he's planning, largely to ensure it all fits. A veteran of small case builds, Stuart cautions other wannabe makers to leave far more room for cables than you think you'll need. He also admits to treating his Raspberry Pi collection rather roughly: "even though they have been abused and tortured, they still keep running."

Stuart assembled the ZX Spectrum Raspberry Pi build from what he had to hand. He took a sheet of scrap metal and used a bandsaw to fashion a crude

▲ The handcrafted heatsink fits beautifully inside the repurposed cassette tape

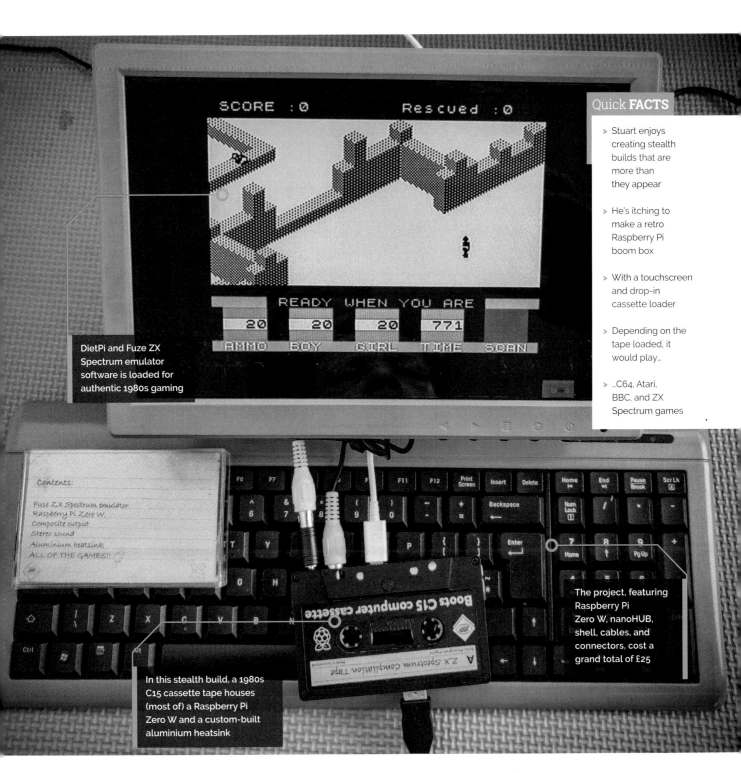

SCORE :0   Rescued :0

READY WHEN YOU ARE

| 20 | 20 | 20 | 771 | |
| AMMO | BOY | GIRL | TIME | SCAN |

Quick **FACTS**

> Stuart enjoys creating stealth builds that are more than they appear

> He's itching to make a retro Raspberry Pi boom box

> With a touchscreen and drop-in cassette loader

> Depending on the tape loaded, it would play...

> ...C64, Atari, BBC, and ZX Spectrum games

DietPi and Fuze ZX Spectrum emulator software is loaded for authentic 1980s gaming

Contents:

Fuze ZX Spectrum emulator
Raspberry Pi Zero W
Composite output
Stereo sound
Aluminium heatsink
ALL OF THE GAMES!!

Boots C15 computer cassette

A ZX Spectrum Compilation Tape

The project, featuring Raspberry Pi Zero W, nanoHUB, shell, cables, and connectors, cost a grand total of £25

In this stealth build, a 1980s C15 cassette tape houses (most of) a Raspberry Pi Zero W and a custom-built aluminium heatsink

shape for what would act as Raspberry Pi Zero W's heatsink. A Dremel, needle files, and fine-grit sandpaper were used to finesse the shape.

Getting it taped

Stuart bought a job lot of cassette tape seconds: "Boots C15 were the cassettes I used for storing my programs back in the '80s; it was an obvious choice" – for which he designed and printed new labels. "Cassette shells make for a great form factor," says Stuart, "I started with a plain black spare cassette shell and used a small hand file and side cutters to remove the plastic supports in preparation for fitting the heatsink."

The 5 mm interior of the C15 cassette tape meant something would have to give: fitting a Zero W

▲ Stuart has form
making retro
gaming builds using
Raspberry Pi

▶ Customised cassette
labels complete the
1980s look

Alert!
Warranty voiding

This project involves
cutting off the GPIO pins
in order to fit Raspberry
Pi Zero W inside a
cassette tape. This risks
damaging Pi Zero W
and invalidates your
warranty. Wrapping it
in aluminium foil while
cutting is a sensible
precaution. Cut carefully.

Make a tape

01 Install DietPi (**dietpi.com**) and the Fuze ZX Spectrum emulator (**fuse-emulator.sourceforge.net**) on your Raspberry Pi and set it to autorun at startup.

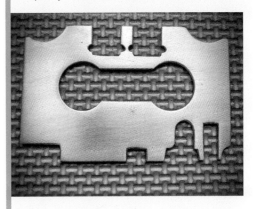

02 Create a heatsink following the contours of the cassette case, and avoiding the spool area. Wrap Raspberry Pi Zero W in aluminium foil and very carefully cut off the GPIO port section where it will prevent the spool wheels fitting.

03 Position Raspberry Pi Zero W in the cassette shell, then fit the ports. Stuart used GPIO sound and an RCA connector for composite video out and added a shutdown button on the front of the tape, and then hooked it up to a shutdown script.

> ❝ I lost some GPIO ports, but it was well worth it to get the tape looking right ❞

▲ Stuart's cassette edition of the ZX Spectrum atop an original Spectrum keyboard

inside involved cutting out a section to nestle under the reels and "preserve the illusion" – not something inexperienced makers are advised to tackle. Stuart has eight similar builds under his belt, hence his confidence. He wrapped Raspberry Pi in foil and "went at it with a Dremel." Surprisingly, it survived. "I lost some GPIO ports, but it was well worth it to get the tape looking right."

Configuring the DietPi and Fuze ZX Spectrum emulator took lots of tweaks before Stuart was able to get them to boot in an acceptable time frame. "I eventually got it to boot in 16 seconds. The full-width heatsink meant I could safely overclock Zero W and saved another couple of seconds," he says.

His next challenge: a 1980s boom box with drop-in cassettes that boot up and play games from different iconic home computers. We like his thinking! ▨

Retro barcode scanner

Forget plug-and-play! Neil Thomas and Chris Harris have created a scan-and-play system as part of a replica retro video game shop, as **David Crookes** explains

MAKER

Neil Thomas

Neil is a retro tech YouTuber who has been running his channel as his full-time job since 2019. His first computer was an Amstrad CPC464!

magpi.cc/
rmccave

▶ Games are loaded using a barcode reader powered by Raspberry Pi, called Barcode Rattler (**magpi.cc/bcrattler**)

F or the past five years, Neil Thomas has indulged his passion for classic technology by presenting a popular YouTube channel called RMC – The Cave. Tens of thousands of viewers watch him bringing the past back to life (literally, in some cases, by repairing old machines). But success ended up creating a small problem for Neil: what to do with all of the retro consoles and computers he's ended up collecting.

"It felt a shame to go to so much effort in restoring and repairing these classic machines only to hide them away in a cupboard," he says. "A pipe dream began to form that I might one day be able to open an exhibition space for retro fans to come and enjoy them and, through hard work

and incredible generosity from viewers, here we are, on the cusp of opening up to the public."

Hey MiSTer

Located on the top floor of an 18th century mill in Chalford in Gloucestershire, The Cave, as the space is called, includes lots of games consoles and computers from the past 40 years, as well as classic arcade machines. But one of the stand-out sections is a replica retro video game store where visitors can pick a game off a shelf, scan its barcode at a kiosk, and quickly start playing – a system that has a Raspberry Pi 3 computer at its heart.

"We'd just created new hardware called the MiSTer Multisystem which was powering the kiosk in the shop, but choosing a game to play on it required me to open the kiosk and manually select one, and that meant there would really only be one playable game all day," Neil recalls. "I wanted to make it more fun with a front-end menu that wouldn't confuse visitors, and then realised the room itself would make the perfect physical front-end. It didn't need to be a menu on the screen!"

At first, Neil and his friend Chris Harris considered using a Raspberry Pi Camera Module to scan the barcodes. "But we found that environmental changes in light and mounting it behind the smoked perspex of the kiosk made it less than 100% reliable," Neil says. "We then switched to a handheld barcode scanner which not only works great but also fits the theme of a retro shop very well."

Space invader

Raspberry Pi runs a program called Barcode Rattler, created by Chris. It processes information from the scanner connected to Raspberry Pi via the USB port, and the information is sent via a secure network connection to the MiSTer gaming

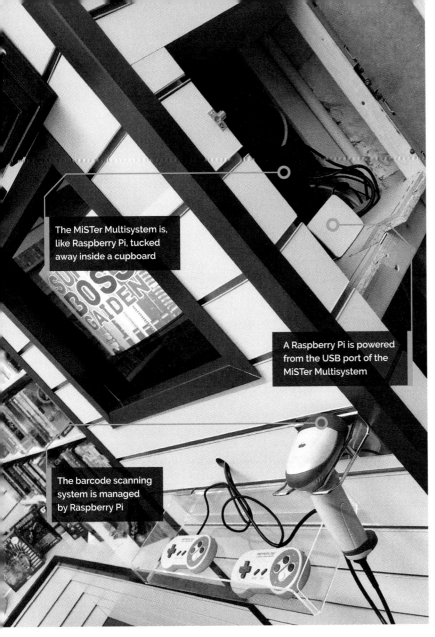

The MiSTer Multisystem is, like Raspberry Pi, tucked away inside a cupboard

A Raspberry Pi is powered from the USB port of the MiSTer Multisystem

The barcode scanning system is managed by Raspberry Pi

Try the latest games

▲ Visitors can just take a game from the shelf and scan at this kiosk which is inside a replica retro video game shop

Quick **FACTS**

> The games play on a CRT monitor

> One button turns the whole system on

> A database of barcode data was created

> The games play via a MiSTer Multisystem

> ROM sizes limit games to 8- and 16-bit titles

device – a highly accurate system well-loved by retro fans that emulates machines using an FPGA chip rather than software.

"Python waits for a keyboard event which it then reads and looks up a barcode in a CSV file," Neil explains. "The barcode scanner behaves as a keyboard device so, when it scans a code, it sends

> ## ❝ People seem to enjoy zapping the games and they love the 'beep' of the barcode gun ❞

the string to Raspberry Pi as if you'd typed out the numbers on the keyboard. If it finds a matching barcode in the CSV, it will send SSH commands to the MiSTer to start the correct system core and load the game via a utility called MiSTer Batch Control."

The result is a fun physical user interface that has gone down well with those who have tested it.

"People seem to enjoy zapping the games and they love the 'beep' of the barcode gun," Neil says. It's also become his favourite room. "It's a time-warp where you can forget the real world for a day and be swept away in a wave of carefree nostalgia," he adds. "What could be more nostalgic than revisiting the video game shop from your childhood?" ▥

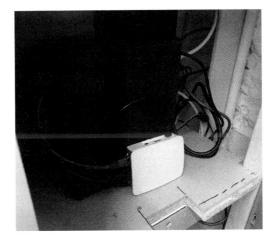

◄ The MiSTer Multisystem was created by Neil in collaboration with the electronics firm Heber, and allows for accurate hardware-based emulation

Mini Pinball Machine

Gamers are sure to have a ball with Chris Dalke's fun desktop pinball table. **David Crookes** has his fingers at the ready...

MAKER

Chris Dalke

Chris is a Boston-based developer and maker. He loves building user interfaces for hardware. At work, he builds autonomous boats.

magpi.cc/
minipinballgit

Alert!
Laser cutting

This project involves laser cutting. Be careful when using a laser cutter in your projects.

magpi.cc/
lasercuttersafety

▲ The laser-cut maple exterior gives a solid arcade feel

Pinball machines have been around for decades. They were popular during the Depression in the 1930s, banned under US gambling laws for 34 years from 1942, developed flippers in 1947 and saw a resurgence in popularity in the 1990s. But, in all that time, the machines have had one thing in common: their sheer size and weight.

That, however, didn't stop Chris Dalke from trying to create a version of his own. "Years ago, I had a high school woodshop class with access to a CNC wood cutter," he says. "I tried to make an electromechanical pinball machine but never finished because the project was too ambitious in scope for my skill set and budget at the time."

Even so, the desire to create a pinball machine remained strong so he tried again, this time creating a miniature version using a Raspberry Pi 4 computer, an Arduino Uno, an LED matrix display, a bunch of buttons and a 7-inch HDMI touchscreen. "It constrained the project to a more realistic scope," he says. "It also allowed the enclosure to be smaller so it could be brought out and played on a tabletop."

Lane change

Chris had a clear objective in mind from the start. "I wanted to retain the feeling of a physical arcade game with intense sound, vibration and colours as well as the tactile response of the inputs," he says. "I noticed many arcade games feel very good at the lowest level of tactile response, an individual button or joystick press, so I started there, with clicky arcade buttons."

To that end, he decided not to replicate the mechanics of a pinball machine. "It was less about the pinball machine and more about building the complete experience of a small arcade game that could retain the feel of a full-scale game," he explains. It led him to create a screen-based version of pinball which he coded in C++ and OpenGL, using the open-source software development library Raylib to create both the graphics and audio.

"I chose very vibrant neon-inspired colours: purple, green and pink which are very saturated on the monitor," he continues. "In the game code I also added extensive juicing which is the use of many small animations and visual/audio tweaks to improve the feel of a game. For example, the

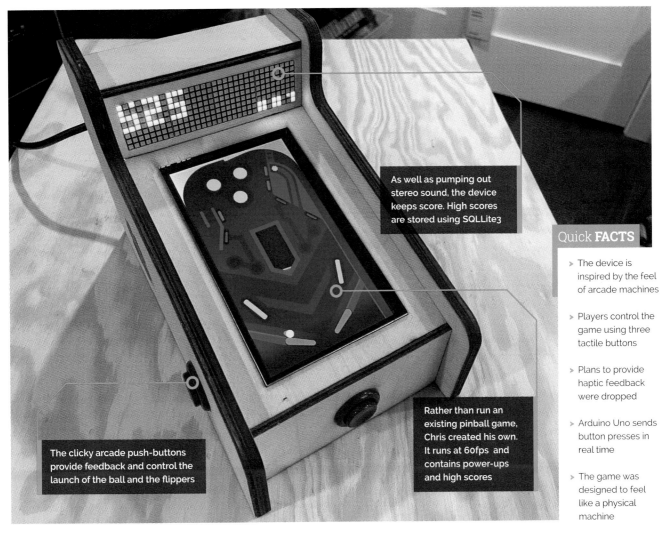

As well as pumping out stereo sound, the device keeps score. High scores are stored using SQLLite3

The clicky arcade push-buttons provide feedback and control the launch of the ball and the flippers

Rather than run an existing pinball game, Chris created his own. It runs at 60fps and contains power-ups and high scores

Quick **FACTS**

> The device is inspired by the feel of arcade machines

> Players control the game using three tactile buttons

> Plans to provide haptic feedback were dropped

> Arduino Uno sends button presses in real time

> The game was designed to feel like a physical machine

🔲 The LED matrix casts a very nice orange light on to the wood 🔳

ball and bumpers stretch and distort excessively when a collision occurs, exaggerating the physical effect of the collision."

Keeping score

For an authentic look, Chris naturally wanted the laser-cut Baltic birch plywood enclosure to resemble a pinball machine, so he tweaked the design to ensure it was unmistakable. "Initially, I'd designed a flat box without the vertical headboard seen in conventional pinball

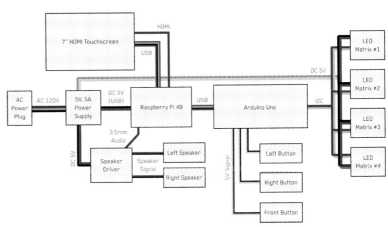

▲ Raspberry Pi 4B and an Arduino Uno are at the heart of the build. Chris says the project was iterative, however, so he doesn't have build plans

▲ Chris told us he loved how arcade games rely on simple, satisfying interaction methods that feel good to play

machines but I added an LED matrix and vertical section," he says. "I wanted to retain the visual signature of a pinball machine and have some element that made the game feel less like it was played only on the touchscreen." YouTube videos such as Secrets of Game Feel and Juice (**magpi.cc/secretsgamefeel**) helped Chris when designing the game.

▲ The top panel of the enclosure holds four Adafruit LED Matrix panels controlled over I2C

> ## ❝ I wanted to retain the feeling of a physical arcade game ❞

The Adafruit LED matrix provides the score and feedback and it certainly looks the part. "The LED matrix casts a very nice orange light onto the wood and it works well to pull the gameplay out of the screen and into the physical world," Chris continues. The Arduino Uno drives the LED matrix and button inputs, and it communicates with the Raspberry Pi board via a serial protocol. The Raspberry Pi is connected to a speaker driver too, allowing for stereo sound.

"I also included a solenoid which I planned to trigger for haptic feedback," he says. "But the vibration was too high-frequency to match the expectation of a heavier ball – I ended up using sound effects instead." Still, this doesn't detract from the overall build, and Chris is very pleased with how it's turned out. "The project is the sum of many individual tweaks to the components, but the whole experience comes together very well." ◾

Creating a Mini Pinball Machine

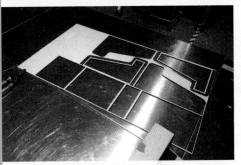

01 After creating a CAD render of the Pinball Machine around the size of a Raspberry Pi 4 and a HDMI screen, Chris laser-cut each of the enclosure pieces out of Baltic birch plywood.

02 Each of the wooden components were designed to interlock so that the enclosure could be assembled with as few screws as possible while also allowing easy access when required.

03 It may look like there's ample room but getting all of the electronics inside the enclosure was tricky. Slats are cut into the side of the enclosure to allow players to hear the stereo sound.

Commodore 64
Pico emulator

Raspberry Pi's tiny but mighty Pico helps bring a classic computer back to life. **Rosie Hattersley** hears more

MAKER

Kevin Vance

A software engineer by trade, Kevin enjoys coding but only recently discovered Raspberry Pi!

magpi.cc/
c64picogit

Despite his job as a software developer, Kevin Vance was new to the world of Raspberry Pi when he began to sketch out plans to revive his old Commodore 64. The retro rebuild of his first ever computer thus became his first Raspberry Pi project. Kevin had been coding since he was young, but only began to immerse himself in the world of digital making as recently as 2020.

Kevin had already started working on making an expansion board for the Commodore when he first learned about Raspberry Pi Pico. "The PIO [Programmable I/O] system sounded really interesting to develop for, and the large flash storage and number of GPIO pins made it ideal for this project, so I started on a new design around it," he explains. The first iteration of the project had a separate microcontroller, flash memory, and a voltage regulator. Kevin was "pretty excited" when he realised Pico included all of those in one module at a much lower cost.

Moving the goalposts

Originally, Kevin wanted to build a Commodore 64 game cartridge with a microcontroller that the Commodore could offload work to. Having followed Ben Eater's "excellent" video walkthroughs on creating a 6502 computer on a breadboard (**eater. net/6502**), Kevin planned a similar scenario with the 6502 machine code stored on a normal EEPROM. The breadboard he designed for his updated Commodore 64 was only his second ever PCB design. Hand-soldering tiny surface-mount

components with a fine-tip soldering iron was "error-prone and required patience," so investing $16 in a hot plate was "money well spent!"

Kevin wrote brand-new code for his project, with frequent updates since he kept changing how the board worked. As he gained a better understanding of how the Pico's PIO and DMA controllers could work together, he decided to investigate whether he could use Pico's RAM instead of an EEPROM. "It

> **"** I wanted to see if I could use the Pico's RAM instead of an EEPROM. It worked better than I expected **"**

worked better than I expected," he tells us. "The PIO state machine could put data from the Pico's RAM on the C64's data bus without involving its CPU, well before the Commodore tried to read it!"

His biggest design challenges were the large number of pins and the Commodore's signal voltage. "Since this is an old machine with a parallel bus, there are not enough GPIO pins on the Pico to hook them all up. Fortunately, since I'm only emulating a ROM cartridge, I can get away with just 14 address lines, eight data lines, and two control lines", explains. He added 5V-tolerant buffers to translate the signals for 3.3V so the Commodore 64's 5V signals would not damage Pico's GPIO pins.

Kevin Vance resurrected his very first computer, using Raspberry Pi Pico's PIO functionality

To get Raspberry Pi to boot into BASIC a switch toggles between an 8K ROM and a 16K ROM

Pico is able to run the Commodore 64 by booting via its expansion port. Using Pico's RAM rather than EEPROM proved a savvy choice

Quick **FACTS**

> Kevin's project rebuilt his first ever computer

> A cheap hotplate helped with soldering accuracy

> He copied the code for the game *Frogger* into Pico's RAM…

> and was astounded when the C64 booted, and *Frogger* loaded

> The whole project cost him around £27 in parts

Pico's PIO system provided the biggest advantage for this build, allowing him to keep all the complexity off the CPU and run with predictable timing.

He used address decoding to send commands from the Commodore so that reading from a special 256-byte block of memory would send the address as a 'command' to Pico's CPU using the RX FIFO. He cleverly factored in time for the commands to be completed by getting the CPU to signal when it was ready for more commands using the TX FIFO. "This lets the Commodore poll the command status without interrupting the Pico's CPU," he explains. "There's a world of interesting devices that this could allow the Commodore to communicate with" – something he plans to investigate further.

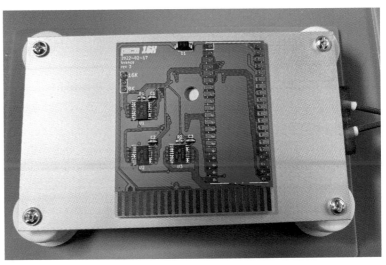

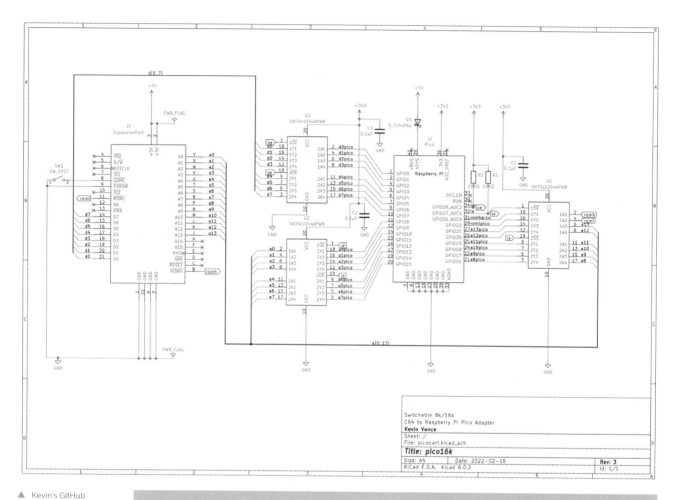

Switchable 8k/16k
C64 to Raspberry Pi Pico Adapter
Kevin Vance
Sheet: /
File: picocart.kicad_sch
Title: pico16k
Size: A4 | Date: 2022-02-16 | Rev: 3
KiCad E.D.A. kicad 6.0.2 | Id: 1/1

▲ Kevin's GitHub page shows a detailed schematic of how Pico is able to communicate with the C64

▶ Kevin was delighted to find *Frogger* loaded up on his revived C64!

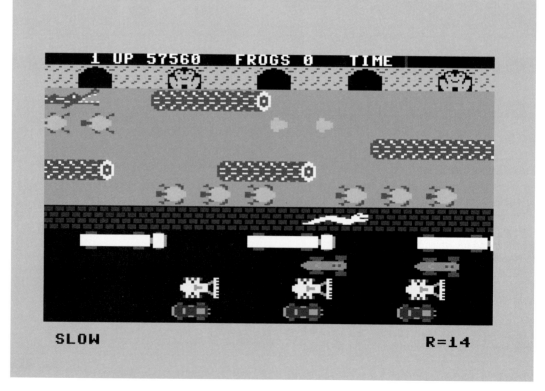

Resurrect a Commodore 64

01 For this project you will use the expansion port on an original C64 and connect Raspberry Pi Pico. Full hardware details and code can be found at **magpi.cc/c64picogit.**

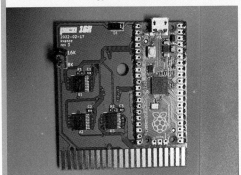

02 Follow the instructions on the GitHub to create a circuit board, and then connect to Raspberry Pi Pico.

03 Fit the resulting hardware into a small case with the expansion port connector slots exposed. Install the provided code on Raspberry Pi Pico.

Extension plans

Although Kevin is pleased to have his childhood computer back up and running, he's still curious about which C64 programs he can use Pico's CPU for – perhaps games he wrote back when his C64 was new and for which he still maintains code – and what devices he can use from the USB port.

He's also become a big fan of Raspberry Pi Pico. "It can do so much per clock cycle and, together with the DMA controller, it can function without the CPU at all once it's been initialised!" He used a second Pico to simulate the C64 bus, speeding up development of the rebuild project. This second Pico may very well become a dedicated microcontroller for automated testing. ▨

▲ Using a second Pico as an emulator helped with the development process

Vectrex Mini

A bespoke version of a rare video games console is within your reach... if you're willing to make one. **Nicola King** learns how to 'mini-fy' an inimitable console of yesteryear

MAKER

Brendan Meharry

Brendan runs a YouTube channel about retro gaming called Retro Game On. As a keen maker too, he tries to mix the two hobbies.

retrogameon.com

Alert!
Copyright

Video game files are protected by copyright law. Be sure to use ROM files that have been released with the owner's blessing, or modern homebrew games designed to be shared. There are lots of legal options.

magpi.cc/legalroms

▶ Testing out the electronics for the controller unit

Remember the Vectrex video games system? **Maybe not.** Launched in 1982, it soon disappeared. Still, it was one of the most curious consoles ever made, with an integrated vector CRT monochrome monitor – colour could be added with the addition of an acrylic screen overlay that came with each cartridge-based game. Only 28 titles were officially released, including the built-in Mine Storm, although some homebrew games were later created by fans.

Anyone now wanting to get their hands on a Vectrex will find themselves hugely challenged, because this unique games system is extremely rare. With that in mind, Aussie retro gamer Brendan Meharry decided to fashion his own version of this classic console, investing some six months of his spare time into the project in the process.

Honey, I shrunk the console

Having experienced the console's delights, playing on an original Vectrex in a local gaming museum, Brendan was motivated to make one. "They were released in Australia, but they're quite rare," he tells us. "I was inspired by official mini consoles like the SNES Mini or PlayStation Classic. I thought it would be fun to 'mini-fy' something more obscure, especially since the original included a built-in monitor." So, with his 3D printer ready to roll, he set to work.

The Vectrex Mini didn't prove too expensive to create, totalling around $70 all-told, not including

▲ Brendan designed the Vectrex Mini case in Blender and 3D-printed it

the spare Raspberry Pi that he already had in his stash. For the audio element of the build, he used a super-small PAM8302 amplifier by Adafruit. Brendan employed a Waveshare 2.5-inch LCD screen for the project, but would he entertain the idea of replacing that with a mini upcycled CRT display? "Maybe! To be honest, the high voltage of CRTs terrifies me (even the small ones), but I'd love to see if someone else could pull it off," he enthuses.

The Waveshare screen plugs into the GPIO pins of a Raspberry Pi 2 Model B which runs a Vectrex emulator using the latest build of RetroPie. "It took me a long time to figure out how to run the emulation in portrait mode with no weird scaling or stretching of the image," says Brendan of one of several challenges he faced along the way.

An achievable goal

Brendan says that building the separate controller, which is based on an Arduino Pro Micro with a KY-023 joystick module, proved to be the easiest part of the build. "If you've watched the [YouTube] video, it's plain to see that it was my first time designing something like that. But on the upside, the later designing/building of the controller went much smoother. It's obvious I learned from my mistakes," he affirms.

Despite the learning curve, Brendan is confident that such a project would be within the grasp of many hobbyists. "The 3D modelling is likely

An LCD is used for the Vectrex Mini's screen, in portrait mode like the original console

Brendan made a separate controller unit for the project

This banana gives you an idea of just how small this minified console is

Quick **FACTS**

> It's powered by a standard, official 5.1 V 2.5 A micro-USB supply

> 3D printing time totalled around 30 hours

> The 3D models are on Thingiverse: **magpi.cc/ vectrexmini3d**

> Brendan has also built a wooden arcade stick running RetroPie....

> View the video for it here: **magpi.cc/ woodarcadestick**

❝ I thought it would be fun to 'mini-fy' something more obscure ❞

the most complex portion of the project, but the wiring, software etc. is quite basic. Overall, I'd say it's a reasonably easy project for an experienced maker."

Brendan has a detailed and informative YouTube video (**magpi.cc/vectrexmini**) taking you through his build, which is well worth a watch if you're interested in fabricating your own version, and he's been bowled over by the response to his project. "As a newbie, I've been blown away by the encouraging comments and constructive criticism," he recounts. But, more than that, he's also gained a huge amount from the experience, and surely that's what we all want to acquire from every project we make. ◾

▲ Final assembly of the 3D-printed case and the innards of the project

Pico
PlayStation
MemCard

Failing hardware and pricey memory cartridges led one maker to use Raspberry Pi Pico to revive his old Sony PlayStation, as **Rosie Hattersley** discovers

MAKER

Daniele Giuliani

Daniele's passion for technology began while tinkering with Raspberry Pi at school in Italy, and he now holds an MSc in Computer Science.

magpi.cc/
picomemcard

Warning!
Cut Safely

This project uses a sharp knife to cut open the original memory card. Watch over children when using sharp knives.

magpi.cc/knifesafety

Software engineer Daniele Giuliani missed playing some of the games on his original PlayStation, but quickly discovered that boxed and unused PS memory cards are hard to find and far more expensive than they used to retail for.

With no guarantee that used official cards – or counterfeit Chinese ones – will perform well, Daniele decided to put his coding skills to good use and create an alternative to Sony memory cards so he could continue playing his old PlayStation games. Having tinkered with Raspberry Pi devices since his high-school days, and subsequently studied computer science at university, Dan realised the platform would be ideal for his MemCard project.

Daniele chose a Raspberry Pi Pico because it's much faster than his beloved Raspberry Pi 1B+, which he still has. A fast GPIO connection was essential for this project. He was also delighted to discover Pico's 'novel' PIO (Programmable Input/Output) interface which he had never seen any board offer. "With PIO you can program, using specific assembly instructions, a set of 'state machines' to control the GPIO directly, leaving the main processor free to do other work," he enthuses. "PIO allows the creation of very powerful bus sniffers. I used it to program the low-level interaction, the basic signals that must be toggled on and off with very specific timings in order to convince the PlayStation into believing an original memory card is present."

Bonus features

Choosing Raspberry Pi for the build brought added benefits: the range of connectivity options improved on what could be done with original memory cards." In particular, since Raspberry Pi can be connected to a PC, it allows [the player] to easily import/export the save files from/to the PlayStation. This is useful to back up old saves and continue playing on emulators," explains Daniele. "PlayStation uses discs to load games. Old discs are full of scratches and can sometimes result in a game freezing in specific parts of the

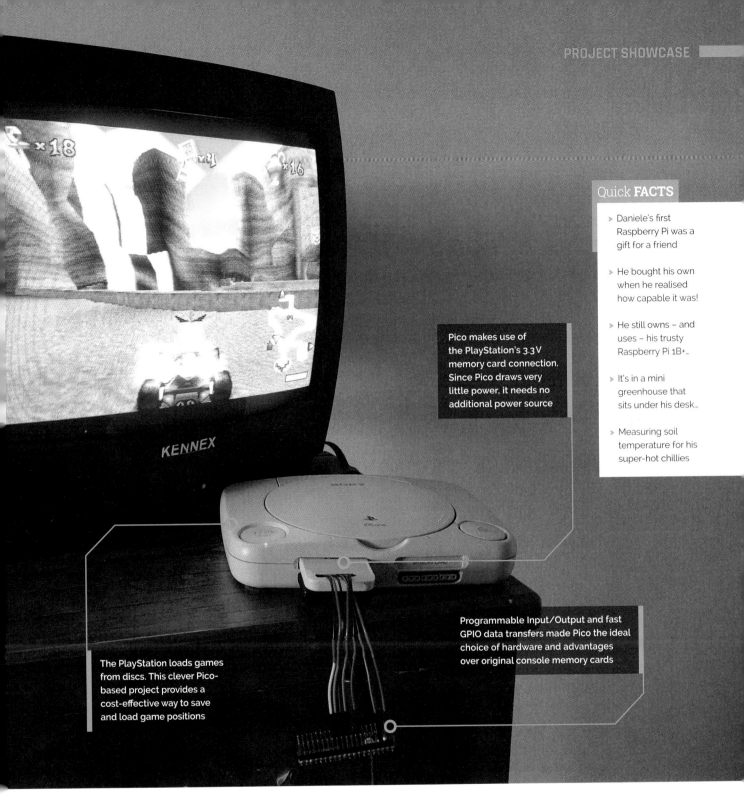

KENNEX

Pico makes use of the PlayStation's 3.3 V memory card connection. Since Pico draws very little power, it needs no additional power source

Quick **FACTS**

> Daniele's first Raspberry Pi was a gift for a friend

> He bought his own when he realised how capable it was!

> He still owns – and uses – his trusty Raspberry Pi 1B+...

> It's in a mini greenhouse that sits under his desk...

> Measuring soil temperature for his super-hot chillies

The PlayStation loads games from discs. This clever Pico-based project provides a cost-effective way to save and load game positions

Programmable Input/Output and fast GPIO data transfers made Pico the ideal choice of hardware and advantages over original console memory cards

game, preventing the user from progressing." With Daniele's Pico MemCard setup, a player can transfer the save file to an emulator, progress through the problematic area (using an intact backup image of the game), then transfer back the new save file and continue playing on the console.

The goal of this project was to provide a solid alternative at a very low price, says Daniele. "Raspberry Pi Pico understands the protocol used to communicate with memory cards and convinces the PlayStation into believing an original memory card is plugged in. Development boards, such as

> Raspberry Pi Pico understands the protocol used to communicate with memory cards and convinces the PlayStation into believing an original memory card is plugged in ▪

▲ The PlayStation
recognises the
Pico memory card
and can save game
data to it

▶ The Scoppy
oscilloscope app
checks for signals
between Pico
and PlayStation

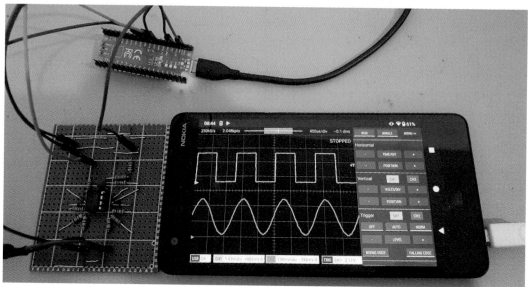

these ones, were particularly interesting because they were quite powerful yet very cheap, and allowed me to set up my personal server to test all sorts of applications."

Smart tools

Although Daniele enjoys the freedom of working on his own projects, lack of equipment was a challenge. A smartphone app called Scoppy (**magpi.cc/scoppy**) proved invaluable: he was able to use a second Pico (plugged into a smartphone) as an oscilloscope to observe the electric signals being exchanged between the PlayStation and the memory card/Raspberry Pi Pico. Details that Daniele found online, of how a PlayStation works, helped him with some of the communication protocols. Prior to finding these details, he'd been trying to read the data between a PlayStation controller and console.

After a month or so of development for his MemCard for PlayStation, Daniele is already planning its next iteration: an enclosure for it, a microSD card version to expand the storage, and support for PS2. He's delighted with the amount of interest that others have shown in his project, and is proud of having found a practical way of reviving a classic, but otherwise unused, games console, potentially keeping them out of landfill. ◩

▲ PlayStation game saves from an emulator can be loaded from the Pico MemCard, enabling players to bypass bad sectors on game discs

Memory card maker

01 The substitute memory card provides a physical interface between the Pico and the PlayStation. Access the electronic board by removing the two screws on the bottom, and take off the plastic shell. Cut a hole in the case for the wires to connect your Pico.

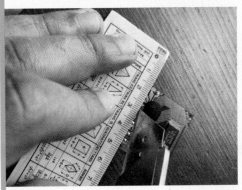

02 Use a Stanley/utility knife to cut a groove right under the pins and disconnect the original circuitry. Use jumper wires to connect your Pico, scraping off their plastic covering, and directly soldering to the copper pins for the PlayStation.

03 Carefully solder your Pico to the PCB, download the latest version of PicoMemcard (**magpi.cc/picomemcardreleases**), plug your device into the computer via USB, and upload the save file.

CRT TV + VCR
Trinitron Retro Media Player

When Mairon Wolniewicz watches movies or plays games from the 1970s, 1980s, and 1990s, he ensures he gets the full retro experience, as **David Crookes** explains

MAKER

Mairon Wolniewicz

Mairon is an Aerospace Engineer from Brazil. While most of his job is coding for satellites, he spends a lot of his free time tweaking configurations and writing scripts as a hobby.

github.com/ MaironW

There is no doubt that a crisp, modern, flat display has been a blessing for our eyes, not to mention the space in our homes, schools, and offices. But if you're a retro enthusiast and you have old video games or movies from the last century, then you may want to go the whole hog and grab yourself a cathode-ray tube (CRT) television instead.

That's exactly what Mairon Wolniewicz has done, creating a media server based around a Raspberry Pi computer that outputs to an old Sony Trinitron CRT he picked up following months of searching. As it happens, these TVs were created in the Sony UK Technology Centre factory in Pencoed, south Wales, where Raspberry Pi is produced today, but Mairon was simply after a quality television set.

"I saw these were the best ever CRTs and one appeared out of nowhere with a built-in VCR," he says. "The seller didn't know if the VCR was working, but I wasn't bothered because my idea was to remove the mechanical parts that occupied the VCR space and use it to store the cables, power supply, and Raspberry Pi."

Picture perfect

In that sense, he had a bold plan. "Since the television only had composite and RF video inputs, I considered connecting the internal wirings of the A/V plug – the ones on the front of the TV – into a Raspberry Pi RCA cable. It would involve cutting out the RCA connectors and soldering the wires directly to the TV," he adds. But then curiosity got the better of him and he bought a copy of *Shrek* on VHS. To his surprise, it worked!

He then resolved to revise his project, deciding, as a lover of retro, that he simply couldn't scrap the VCR element of the TV. Instead, he opted to place the Raspberry Pi 4 computer into a case that he clipped on to the grid at the back of the TV. He admits it's not the most elegant of solutions ("the composite inputs are on the front of the TV," he laments), but his retro media player could still make use of the CRT display.

So what has motivated him to do this? "Well, I've been spending a lot of time learning how to paint by watching videos by Bob Ross [an American painter and TV host who died in 1995], and I thought it would be cool to just let videos like these play in a loop on a small screen in my room," he explains. "Soon, I was also looking to add other functions. My Raspberry Pi came with SNES controllers, so playing games was inevitable."

Play it again

This is when the project took off. The setup was straightforward (videos and games are stored

The original plan was to place the Raspberry Pi computer within a stripped-out VCR space, but this method allows it to be easily unclipped and plugged into an HDMI monitor for debugging

Most of the work has gone into the creation of the retro media system – Raspberry Pi simply plugs into the front of the TV via the composite video output

Mairon uses a Bluetooth keyboard for accessing the menus and inputting, and either Bluetooth or USB controllers for playing games

To keep Raspberry Pi in place and easily accessible, Mairon has made good use of the grill on the back of the TV

❝ I saw these were the best ever CRTs and one appeared out of nowhere with a built-in VCR ❞

on a USB stick and video output is via a P3/RCA connector), but Mairon's retro media system has a custom GUI resembling a VCR interface – complete with white text on a blue background.

▲ Mairon's *Shrek* physical VHS tape playing perfectly on the TV – the reason why he didn't open up his TV (warning: don't forget that working inside old CRT televisions is dangerous)

▲ It's not quite as polished as Kodi, but that's deliberate. It looks like Mairon loves 1980s horror flicks!

ALERT! CRT

CRT monitors and televisions can carry lethal amounts of electrical charge. Be extremely careful when working with CRT-based technology.

magpi.cc/crt

▲ Mairon had been a long time user of RetroPie and EmulationStation, so decided to stick with it rather than create his own emulator/game selector

01 The VCR-inspired menu lets you navigate up and down to watch TV, play video games, or listen to music. 'Exit' was added because Mairon got bored of typing `sudo shutdown -h now`

02 As well as TV shows such as Bob Ross's *The Joy of Painting*, Mairon can navigate to music videos, other series and movies, and random YouTube videos via **MyRetroTVs.com**.

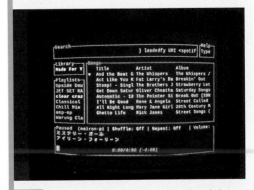

03 Homebrew retro games can be played via EmulationStation and the radio option launches Spotify TUI, a tool that allows Spotify streams to be controlled from the terminal.

"I thought it would be cool to have a GUI where I'd choose what I wanted to do with my media player, and I thought the VCR look would match nicely," he says. This was accomplished using Qt QML, a declarative language for designing user

▲ This retro media system is fundamentally designed to be plug-and-play, with the software used to launch the correct applications

❝ I thought it would be cool to just let videos play in a loop on a small screen ❞

interfaces. "C++ was used for file reading and to launch Linux commands," he adds.

Most of Mairon's time has been spent tweaking Raspberry Pi configuration files. "It will run fine with composite output, on the desired resolution with legible font and without overscan," he says. Indeed, the system is configured to output video to 480i, and it's set up to correctly run music and games.

It's navigated using the up/down arrow and **ENTER** keys on a keyboard, with the **BACKSPACE** key for returning to a previous menu. "I'd like the GUI to support gamepads, and I'd also like to integrate a personal assistant," he says. "Sure, it's not even close to retro, but wouldn't it be cool to request your TV to play a season of an old show when you lie down on your bed?" ▥

Team Pinball

The thrill of pinball, plus the chance to plot a brand-new game, came together nicely with the help of Raspberry Pi. **Rosie Hattersley** hears how

MAKER

János Kiss, Otília Pasaréti, and Romain Fontaine

Team Pinball was founded in 2016 by pinball enthusiasts and open-source advocates János, Otilia, and Romain.

teampinball.com

Team Pinball is the UK's only pinball design-and-build company. Based in Wales – pretty near to the Raspberry Pi factory, by coincidence rather than design – the trio behind Team Pinball come from Hungary and France, and are firm believers in all things open-source. Having seen the success of RetroPie gaming and how well it uses the capabilities of Raspberry Pi, the founders set about designing a new generation of pinball machines.

"Raspberry Pi is a powerful tool that has already found its place in the video game community with RetroPie. We wanted to do the same and create new, fun games with the Raspberry Pi," explains Romain Fontaine. When the team sat down to decide on the theme and title for their pinball

> ❝ The pinball table brings the player back to Chicago and the 1930s Prohibition Era with gangsters, casinos, and a bank to rob ❞

machine, it became clear they were all fond of games from the 1980s era. Their retro game, The Mafia, "brings the player back to Chicago and the 1930s Prohibition Era with gangsters, casinos, and of course, a bank to rob!"

Pi-eyed

Team Pinball originally designed their game for Raspberry Pi 2, noting its compact size and price and that it had "everything we needed", including being able to output audio, drive the screen, and control the machine through its

Their friend Attila Szabo created the striking graphics, while Balint Ats's simulation helped check the gameplay worked as it should

GPIO," says Romain. Having been updated and upgraded in the five years since Mafia Pinball launched to some acclaim in the pinball enthusiast arena, The Mafia is now based on a Raspberry Pi 3 that runs game code on top of a modified Linux image optimised for fast boot and low latency. There are two separate sets of software. One runs the game logic that reads the switches located in the machine and drives the solenoids and light show; the other plays music and audio effects, and also renders the score and animations on the LCD screen.

Starting from scratch

Team Pinball designed the game and cabinet themselves, basing it on their own sketches, cutting playfields, wiring harnesses, and the electronic controller. The game's pinball layout is their own design too, while most of the pinball-specific parts, such as mechanisms, are standard and can be bought on pinball websites. As far as possible, parts were locally sourced: the wooden

Raspberry Pi's SPI bus powers the pinball machines' two lamps and two WS2811 RGB LED chains

Raspberry Pi 3B scans the switches every 3 ms, processes the game logic, and drives the solenoids through shift registers and IRL540 MOSFETs

Quick FACTS

> The Mafia Pinball game took two years to develop

> The first prototype machine cost around £10,000

> A virtual simulator was used to check the layout and gameplay

> Issues included metric-manufactured European parts and imperial pinball parts

> And a 32-tonne truck delivering them all to their rural office

▲ Final assembly complete, Romain applies the Team Pinball logo

▲ Team Pinball commissioned Balint Ats to create the distinctive graphics and decals for the game

◄ Flippers, lights, and other standard pinball parts provide a familiar gaming experience

cabinet was made by a Welsh company, with printing and glass panels also done in Wales. "Our office is only 30 minutes away from Raspberry Pi's factory," says Romain, so they can also count that as a locally sourced part.

Team Pinball designed their own controller board, called Rboard, specifically for Raspberry Pi that is compatible with direct switches, as well as a switch matrix with up to 100 switch inputs. "It can drive more than 200 LEDs and 48 solenoids. This is more than previous pinball controllers such as the WPC, while using one tenth of the board space thanks to Raspberry Pi," says Romain. "Several revisions" to the Rboard added a watchdog circuit to protect the hardware, capacitors, "and a regulator to maintain a perfect 5V for Raspberry Pi." The latest version includes a DAC for improved audio quality. Video needed to be in H.264 format, so they had to find alternative rendering techniques.

Manufacturing was measured in a combination of metric for bespoke parts and imperial for standard pinball parts, which added to the challenge – not least because some of the manufacturers had never worked on a pinball machine before. When the 32-tonne lorry

With commissions to create machines for others, each machine is handmade by Team Pinball

delivering everything to Team Pinball's HQ arrived, turning down the narrow Welsh lane was a feat too far, so Romain and his colleagues had to carry the cabinets for the last part of their journey in the pouring rain. Nonetheless, everything was assembled and comprehensively checked before Mafia Pinball eventually made its much-anticipated debut.

The game has been a big hit in the pinball community, gaining numerous plaudits and orders. Their machines have shipped globally, to USA, Canada, Sweden, France, Austria, and even Australia. A Canadian company also uses the Team Pinball system for their game design. Better yet, Romain says Team Pinball "absolutely!" has more pinball machines in the pipeline. "More pinball projects, all powered with Raspberry Pi, CM4, and Raspberry Pi Pico!"

Get pinning

01 If you want to make your own pinball machine, there are plenty of options. Team Pinball's controller is open-source, and can be found on GitHub: **magpi.cc/rboardgit**. Romain also suggests **pinballmakers.com** as a good place to start.

02 A good knowledge of Python is useful for coding your own pinball machine. Alternatively, you could use code from **missionpinball.org** to either run software on an old pinball machine or for a brand-new one.

03 Low latency to ensure rapid flipper response is vital. Along with a lightweight Linux image, Team Pinball uses C/C++ "to communicate with the GPIO pins for the best performance."

Game Boy Interceptor

A challenge by fellow Game Boy Tetris fans led to this Pico-based recording device, discovers **Rosie Hattersley**

MAKER

Sebastian Staacks

Sebastian is a physicist and developer of the app 'phyphox' at the RWTH Aachen University (Germany). He is a father of two and presents his hobby projects in his blog...

there.oughta.be

The addictive qualities of the computer game Tetris are well-known, so it's perhaps no surprise to learn that there is a whole community of Game Boy Tetris fans who meet online for tournaments.

The RP2040 microcontroller-based Game Boy Interceptor came about when just such a tournament was being planned, "and, of course, they wanted to stream the contestants' gameplay," relates fellow Tetris fan Sebastian Staacks. "Streaming would not be a problem with a modified Game Boy or a modern Game Boy clone such as the Analogue Pocket," says Sebastian, "but it would mean contestants would be forced to use the same platform in order to compete." This change just wouldn't fly: "the contestants always played their favourite Game Boy model and, in a contest, would want to use the model on which they trained their muscle memory."

Getting everyone to modify their beloved handheld console was out of the question. Sensing a challenge he could relish, Sebastian agreed to work out a way of streaming the tournament that would satisfy everyone. His idea was to insert a device between the game cartridge and the Game Boy that checks what the handheld console is doing and reconstructs the gameplay from the data.

Game on!

Physicist Sebastian holds a PhD in solid-state physics, "which means that I know the basics of almost everything technical, but nothing that is required to apply it," he announces modestly. A self-taught programmer and electronics enthusiast since childhood, Sebastian follows his own advice that the best way to learn new skills is "to become obsessed with a project that is just a little bit above your current abilities."

When he got his first Raspberry Pi a decade ago, he wasn't quite sure what he would do with it. Sebastian has been running a Raspberry Pi-based home automation setup ever since, and also keeps a Raspberry Pi handy for the constant stream of projects that invariably need a simple server for IoT purposes. When Raspberry Pi Pico launched, he was curious about Raspberry Pi microcontrollers, especially after working with Arduinos for a while. Happily, "the RP2040 turned out to be a perfect match for GB Interceptor!" Unlike modern devices, the Game Boy reads data from the cartridge as fast as from its RAM, so there is no reason for it to load its code into RAM first. Instead, the code is executed directly from the cartridge (with few exceptions) and a device in between would know exactly what the Game Boy was doing.

Unexpected benefits

At first, Sebastian simply hoped to use RP2040 to capture information from Tetris, but the microcontroller was powerful enough to render the graphics, emulate the code from whichever game was being played, and act as a general-purpose video capture device. The power and sophistication of the RP2040 meant the Game Boy Interceptor was a much more useful and flexible device than Sebastian had anticipated. "You can simply plug it between the cartridge and your Game Boy and connect it to a host device via USB." There, it shows up like a webcam and, as a USB Video

v4l2:///dev/video4 - VLC media player

Sebastian realised that the entire game data would pass through the Game Boy's cartridge slot – and a powerful microcontroller could emulate it

The GB Interceptor acts as a video-capture device for older Game Boys, streaming moves during online Tetris tournaments

Raspberry Pi RP2040's GPIO pins and programmable IOs capture video RAM data and emulate moves without needing the gaming device to be modified

Quick **FACTS**

> The Game Boy has a 1MHz bus...

> ...and uses USB 1.1, with a frame rate of just 29 fps

> So, Sebastian had to be creative in replicating its effects

> Sebastian's LED Cube featured in *The MagPi* #100

> It has since been converted into an artificial fireplace!

▲ Sebastian's GB Interceptor board is based on Raspberry Pi's minimal hardware spec for the Pico, with added USB-C

Capture the gaming action

01 The Interceptor works "like an emulator on rails" and executes the same instructions that the Game Boy receives from the cartridge to reproduce what the Game Boy writes into its video RAM.

02 Snug inside a 3D-printed case, the RP2040 emulates the Game Boy's graphics unit and renders images. "It works almost perfectly in 98% of the games tested," says Sebastian.

03 However, Sebastian had to overclock the RP2040 to 250MHz to get it to work. The Interceptor is currently limited to classical Game Boy games since Game Boy Color games use twice the clock speed.

▲ The GB Interceptor can even be used as a webcam, as Sebastian demonstrates

Class device, it does not need a driver and just works on Linux, Windows, and Android. It works with macOS too, though Sebastian says he still experiences some issues with M1- and M2-based Macs. It can even make use of the Game Boy's camera and function as a webcam.

" The design and implementation came together quickly "

Sebastian had previously worked on a cartridge using the ESP8266 microcontroller, which had the benefit of Wi-Fi, but was far too underpowered to deal with the Game Boy's 1MHz bus speed (**magpi.cc/wifigbcart**). He had also undertaken a Raspberry Pi Pico project using its programmable IOs. Using RP2040 made sense since he didn't need wireless connectivity for the Game Boy Interceptor, whereas the PIOs were adept at reading bus data leaving the CPU free to pick up bus events at its own pace. This, plus the fact he uses Raspberry Pi "a lot", meant the design and implementation came together quickly. The sole changes from his first version were to correct the orientation of some of the LEDs and a switch to USB-C which is useful for indicating whether the Game Boy is on or off. **M**

■ Learn coding ■
■ Discover how computers work ■
■ Build amazing things! ■

5th Edition
Fully updated for
Raspberry Pi 5

The Official
Raspberry Pi
Beginner's Guide

How to use your new computer

Gareth Halfacree

magpi.cc/beginnersguide

MCM/70 Reproduction

Reproducing vintage technology is a labour of love for this maker who is uncompromising in his attention to detail, impressing **Nicola King**

MAKER

Michael Gardi

A retired software developer, living in Waterloo, Ontario, Canada with his wife, who appreciates having the time to make whatever the heck he damn well feels like!

magpi.cc/
mgardiyt

Michael Gardi's reproductions of classic computers have featured in *The MagPi*'s pages on several occasions. A proud Canadian, he enjoys spotlighting "Canada's contributions to the personal computer revolution."

Confident that he "had a pretty good handle on all of the machines available in the 1960s and 1970s", he was somewhat flabbergasted when he came across the MCM/70, a "beautiful and innovative" Canadian-built personal computer that he'd never heard of. Even more so because it was conceived and made in Kingston, Ontario, very near to where Michael lives.

Hardware challenges

Keen to build his own repro version, Michael began to source components. While obtaining a keyboard was straightforward, from Dave at **osiweb.org**, he wasn't sure how to reproduce the MCM/70's unique APL keycaps. After considering making his own, he discovered a helpful UK supplier of custom keycaps, **kromekeycaps.com**.

Michael used Inkscape to create SVG versions of the legends that could be scaled to fit the keycaps. Detail is crucial in his reproductions, as he reflects: "With the 'inside' of the machine being emulated, it is very important to me for the outside to look as much as possible like the original. I want my reproductions to be clearly recognised, and operate the same as the originals they replicate."

Michael spent some time looking for a suitable replacement for the original MCM/70's plasma display that's no longer manufactured, and eventually found the high-performance Broadcom HCMS-2972 dot matrix display which offered a close match. "Packaged as eight 5×7 dot matrix arrays, these modules operate at a nice safe 3.3V and can be cascaded together side by side to create the 32×1 character display desired here."

The aesthetically pleasing casing took Michael many hours to 3D-print, and he added two placebo cassette decks for an authentic look: "Virtually all of the online images of the MCM/70 feature the two tape deck model."

Retro Raspberry Pi

Following a research visit to the York University Computer Museum (YUCoM), Toronto, Michael decided to use the YUCoM working MCM/70 software emulator in his project, which has "high historical accuracy" – important if his version was

> ❝ It is very important to me for the outside to look as much as possible like the original ❞

to work just like the original. "Thanks to the hard working folks at York University, I have a great head start with this project."

To complement that, he needed a solution with a fairly powerful CPU to run the emulator. After first trying it on a Raspberry Pi 2 he had to hand, he found it only executed at about 33% the speed of an original MCM/70. "When I moved the project to a Raspberry Pi 4B, the emulator ran at twice the speed of the original," he reveals.

Raspberry Pi's large number of GPIO pins also allowed Michael to wire the emulator to interact with the display, keyboard, and cassette deck. Plus, he needed "a target system with a Linux-based OS to build and run the emulator (which was written in C). In fact, the emulator built easily on Raspberry Pi OS after a couple of required libraries were loaded."

> Michael worked on the MCM/70 for about a year

> He estimates that it cost him around $750 CAD to create

> The eight separate case pieces each took 8–10 hours to 3D-print

> The cassette desks/display frame took an additional 25 hours to print

> The project is powered by a standard 5V 3A Raspberry Pi PSU

In place of the original machine's discontinued plasma display, Michael used a dot matrix display

Although the tape decks don't actually work, "you can slide a cassette into them and use the release lever to eject the cassettes"

Custom keycaps were sourced and SVG versions of the original key symbols created to label them

The result of his hard work is impressive and Michael has already put it to good use. "I've been using the MCM/70 Reproduction to learn APL by working my way through the MCM/70 User's Guide. Mind you, I won't be writing any APL programs any time soon, but I would like to get a good feel for the language."

He has also generously shared a detailed description of his project on Instructables (**magpi.cc/mcm70repro**). It's well worth a read, illustrating his enthusiasm for his retro subject, and taking you through how you can create your own version. ▥

▲ Held in a 3D-printed caddy (bottom right), Raspberry Pi 4 is connected to the dot matrix display, and to the keyboard via an encoder board

RAD Expansion Unit

Want to get the most out of old tech? **David Crookes** looks at one device which boosts the Commodore 64 so much, it's capable of playing Doom

MAKER

Carsten

Carsten is a computer graphics researcher and lecturer. Since rediscovering computers of his childhood, he tries to convince Raspberry Pi boards to communicate with them.

magpi.cc/radeu

▼ Carsten has created a number of different expansion units for the Commodore 64, all of which can plug into the computer's expansion port

Considering Doom was once thought impossible on the Commodore Amiga, running it on an 8-bit Commodore 64 would appear to be sheer fantasy. But, thanks to the efforts of maker Carsten, such an achievement has become reality.

Having previously created a multifunctional cartridge called Sidekick64 for various Commodore 8-bit computers which emulated memory expansions, sound devices, freezer carts and much more, Carsten saw that many people in the retro community still wanted a replacement for the official RAM Expansion Unit (REU) which was released in 1986 and discontinued four years later.

For those unaware, the REU plugged into the C64's expansion port and added extra memory while also allowing for direct memory access (DMA) transfers – something the Sidekick64 couldn't do. This meant data could be transferred to and from the main system memory whole bypassing the MOS Technology 6510/8500 CPU. "My RAD Expansion Unit was designed to do these transfers," Carsten says. It's a major triumph!

Radical thinking

There are many benefits to having the RAD Expansion Unit, which boosts the amount of

available memory from the standard 64kB to as much as 16MB. "It helps to run Geos [a C64 OS] and it makes some tasks less annoying, such as copying disks in one go. It also functions as a RAM disk to accelerate working with the system," Carsten says.

In order to create the project, then, two PCBs were produced: one to fit Raspberry Pi 3A+/3B+ and another to fit Raspberry Pi Zero 2 W (the latter creating a less expensive unit). "I didn't want the glue logic that facilitates the bus communication with the available GPIOs to get too extensive and I wanted to avoid Complex Programmable Logic Devices and such which would prevent many people from building their own RAD," Carsten says.

Indeed, the project was designed so that the RAD contains the glue logic to interface Raspberry Pi with the C64 bus. "You simply put it on to Raspberry Pi like a HAT and plug it into the expansion port of the Commodore 64," Carsten continues. "The combination of a fast SoC and a decent number of GPIOs was great."

An explosive result

There were challenges, of course. "The biggest was getting the bit banging right," Carsten says. "Most of the communication has to happen within a time window of less than 500 nanoseconds – most often there's significantly less time between all signals being read and putting data on the bus."

To make the device run smoothly, Carsten had to use multiplexers. "More signals on the expansion port need to be read/written to than Raspberry Pi has GPIOs," he explains. The correct data had to be put on the bus at the right time to prevent memory corruption and, in the worst case, random instructions being executed by the CPU, causing a crash. "In general, to get the timing right, I needed to hit intervals at a spacing of approximately tens of nanoseconds, which I did using CPU cycle counters."

With all that in place, it was time to chill and RAD-Doom proved a great way to do so. Most of the processing is being done by Raspberry Pi ("it's essentially a CPU replacement where the new CPU

The device has allowed Doom to run on a Commodore 64 or Commodore 128 at a stable 50 frames per second

Press the menu button on the RAD and it's possible to manage or browse images, apps, games, and Nuvie videos

The device plugs into the C64. Raspberry Pi boots a bare metal kernel from the SD card which handles all the communication and functionality

Quick **FACTS**

> The device emulates RAM expansion

> It also provides a CPU boost for the C64

> RAD uses DMA transfers

> The C64's graphics chip is still used

> You can buy the device ready-made

is a one-core ARM running at 1.4GHz with its own 512MB RAM," Carsten says). But the important thing is that the tech demo uses the C64's VIC-II graphics chip – and works! Sound is also streamed to the iconic SID chip.

❝ It makes some tasks less annoying, such as copying disks in one go ❞

"I wanted to see how the C64/C128's VIC-II and SID performed if CPU power and memory was not an issue," Carsten says. "I also wanted to experiment with real-time colour dithering for the VIC-II." By making use of Doom to output graphics and sound, he's certainly managed to achieve that. ⚠

▲ If you want to make your own RAD, you can download and print a case for it too

Pico Held

David Crookes takes a look at a device based around Raspberry Pi Pico that emulates the Sega Mega Drive and lets you play homebrew games

MAKER

Daniel Kammer

Daniel is a physicist working as a teacher in Germany. He loves embedded and retro systems because he enjoys the challenge posed by limited resources, as well as their huge DIY potential.

magpi.cc/pplib

Although there are many handheld retro game consoles based around Raspberry Pi devices, we still want to play with more. They help to keep the memories of old machines and titles alive, while providing the perfect opportunity to dive into a project and learn more about what is going on under-the-hood.

It's this desire to understand that has driven Daniel Kammer to create Pico Held, a beautiful, open-source handheld with a Raspberry Pi Pico board at its heart. "There is so much you can learn from creating DIY projects," he says. "By making it open-source, it also gives others the opportunities to learn – maybe they'll help to improve Pico Held in the future."

Hold fire

Before getting down to work, Daniel came up with some goals. "I wanted an appealing design, a screen big enough to play games on, a nice

The screen is connected using a parallel interface which offers transfer times up to eight times faster than an SPI connection

speaker for sounds and music, and a device powered by Raspberry Pi Pico – an amazing piece of hardware with a balanced set of features," he explains.

There were clear reasons for this. "A small screen and subpar speaker kind of defeats the whole point of having a gaming device in the first place," he notes. This prompted Daniel to hunt for the largest, low-cost screen possible. "A 3.2-inch display was the biggest IL19341 I could get on AliExpress," he says, but the goals posed a host of challenges.

Since Daniel custom-made a PCB and 3D-printed a case, he needed everything to fit. "I had to align the button holes to the PCB contacts, make sure the LCD was nicely centred, and the analogue stick was fitting," he explains. "I also found it quite challenging not to exceed a thickness of 12 mm. For every flaw, I made a new PCB and new case!"

In the end, sacrifices needed to be made. "My prototype had four buttons, but I decided to go for three and use an analogue stick instead of a D-pad," he says. "I also wanted to design the Pico Held in a way that others could build it too. That was a big constraint."

To be this good

Developing the software proved even more challenging. Daniel wanted his handheld to play retro games, so he modified a Sega Mega Drive

▲ Daniel says homebrew game developers would be able to find a way to cope with three buttons. "DIY is about creativity, right?", he says

As well as a 3 W speaker, there's an analogue control stick, three buttons, and an on/off switch

The 3.2-inch screen has a resolution of 320 × 240 pixels and it sits in a case that's 136 × 60 × 12 mm

emulator called Gwenesis by bzhxx. "I had to fiddle with the endianness [the order in which bytes are stored] and the RAM/ROM handling," he recalls. "It was pretty frustrating because virtually all errors were hard faults."

The result has been an emulator running Mega Drive games more slowly than intended, but Daniel says lots of games still remain playable.

> ## A small screen defeats the whole point of having a gaming device in the first place

"Sound was a bit trickier: I had to make the sound generation per frame/per line rather than per cycle, and that causes the sound to be flawed in a lot of games." Daniel has also been able to run a NES emulator on his device: "It runs at full speed," he tells us.

More than that, Daniel wants people to create their own games. "The Pico Held's software library was written with the idea of allowing people to create classic pixel art games with 256 colours," he reveals. "Creating the blitter and the tile map blitter in the software library for this purpose kept me busy for days."

Yet, it's been worth the effort. "Some have said the analogue stick isn't well-suited for this kind of game device, and some want four buttons," he says, confessing that he isn't actually really into gaming. "But the reactions have been surprisingly positive. I was really glad about that!" ◾

◀ Raspberry Pi Pico works well as a console emulator. RP2040 Plus can be used instead to provide 4 or 16 MB of flash memory

MAKE YOUR OWN GAMES

PROGRAM RETRO-STYLE GAMES WITH RASPBERRY PI

...You just don't g
DEMAND to see t

State of Play:
Game Engines
for Raspberry Pi

Develop anything from text adventures to 3D
worlds with four sophisticated game engines

MAKER

**K.G.
Orphanides**

K.G. is a software
preservationist and
developer who's
written middling
text adventures
for fun and done
despicable things
to 3D engines
for profit.

magpi.cc/owlbear

You don't need a fancy integrated
development environment (IDE) or
graphical engine to create fantastic
computer games, but they really can help. With
mature choices of 64-bit operating system for
Raspberry Pi and enthusiasm among developers
for releasing software to support Linux on
aarch64 hardware like Raspberry Pi 4's, game
engine support for the platform has expanded
significantly since we last explored our options
for game development. We're going to highlight
four of the most exciting game engines you
can use to make your own games on Raspberry
Pi today.

Godot 3.5

magpi.cc/godot35

64-bit? Yes
32-bit? Yes

Godot is a free, open-source game engine for 2D and 3D
games. In fact, it's one of your few fully functional choices
for 3D games development on Raspberry Pi, and includes
a full IDE in the vein of Unity and Unreal Engine. That said,
the editor does chug a bit with a reasonably busy 3D
scene loaded in, so you might want to stick to 2D, or only
the simplest 3D scenes, if Raspberry Pi is your primary
development system.

We recommend the Unofficial Godot for Raspberry Pi
port. On the downside, this means you're limited to
Godot 3, rather than the slightly snazzier Godot 4. The
latter can also be built for Raspberry Pi, but there isn't
currently a consistent deployment path for this, so it's
best to stick with the Unofficial Godot port of 3.5 for
stability. Both Godot 4 and the Raspberry Pi port of Godot
3.5 allow you to target your game builds at Raspberry Pi.

By default, Godot uses its own GDScript language, a
high-level, object-oriented language with Python-like
syntax. It's very quick to learn and is, in fact, your only
option in this Raspberry Pi port, although other versions
of Godot offer optional support for C#, C, and C++. Godot
has comprehensive documentation with some fantastic
tutorials to get you started – just make sure you're
following the docs (**magpi.cc/godotdocs**) for version 3.5.

Install it:
Download the latest release for your Raspberry Pi from
magpi.cc/godotrpi.
Unzip it and run the relevant version.

Notable works:
Brotato, Cassette Beasts, Cruelty Squad, RPG in a Box.

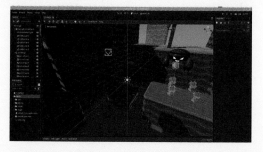

▲ Godot is a full 2D/3D game engine and integrated
development environment. Everything works, though
3D scenes can run slow when you play them

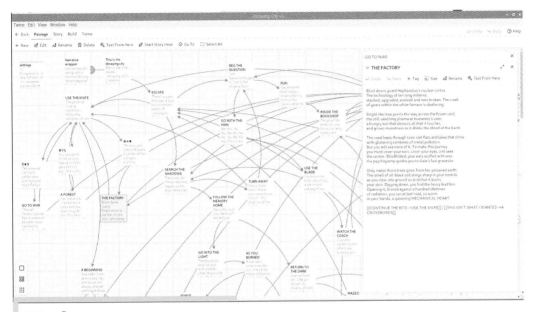

Twine allows you to create complex choice-based games. Minimal programming is required, but you can create surprisingly complex routines using story formats like Harlowe

Twine

twinery.org

64-bit? Yes
32-bit? No

Twine is one of the most approachable ways of making a text game, allowing you to create choice-based interactive fiction and choose-your-own-adventure narratives. It's based on HTML and JavaScript, making it easy to publish your games online and even extend the features of the game engine, which by default supports sound, images and video, just as HTML5 does.

Twine has made its mark in the world of interactive fiction over the years, has become one of the most popular engines in the IFComp contest, been used by narrative designers and game writers as a prototyping and plotting tool in larger projects, and had a significant presence in the LGBTQ+ games movement of the 2010s. We've been running Twine on Raspberry Pi for years, starting with Raspberry Pi 3 running a 32-bit operating system. Things have moved on since our 2016 tutorial, way back in issue 64. Although our guide to developing for Twine's Harlowe language is still relevant to modern versions, Twine's IDE has seen some welcome polish in recent years.

There are four built-in story formats: Chapbook, Harlowe, Snowman, and SugarCube. Your choice of story format determines the appearance, behaviour, and syntax of your Twine creations. Harlowe is the beginner-friendly default format and our recommendation for anyone starting with Twine, as it's both easy to use and flexible enough to allow you to easily bolt on additional functionality.

SugarCube is the continuation of Twine's first story format, and benefits from a mature macro language; Chapbook is a more 'literary' and generally book-like story format, great for making traditional choice-based books, and good for authors without any programming knowledge. Snowman is a bare-bones but highly customisable story format designed for developers who are already comfortable with JavaScript and CSS.

Whichever story format you choose, Twine's development interface presents you with a series of interlinked boxes, each containing a passage of text that you can think of as a paragraph, a page, or a location in your adventure game. Your finished game will be spat out as an HTML5 plus JavaScript experience that can be played via a web browser – just upload it to your website and share the link. A number of tools and wrapper techniques exist to turn your web-based game into a stand-alone offline experience, although that's not the primary focus of the engine. Indie game distribution platform itch.io directly supports Twine games.

Install it:
Raspberry Pi users running a 64-bit operating system should go to **twinery.org**, click 'Download desktop app', select the Linux-arm64 version, unzip it, and run the file named **twine** from a terminal with `./twine`.

Notable works:
Depression Quest, This Book is a Dungeon, Horse Master: The Game of Horse Mastery, Black Mirror: Bandersnatch (development).

magpi.cc/inform

Inform 7 is a mainstay of the creative interactive fiction community and has also been used for commercial titles such as Andrew Plotkin's Hadean Lands

Inform 7

64-bit? Yes
32-bit? No

Inform is one of the best-known interactive fiction languages, and version 7 is the pinnacle of its development: a natural-language text adventure authoring system in English; to create a space for players to interact with, you type something like 'The Shed is a room,' followed by its description. Interactive objects and characters are defined in a similar way. Open-sourced in 2022, Inform 7 allows writers to use the familiar milieu of words to develop games with complex functions and rich interactivity.

The Inform 7 IDE is an immensely useful tool. It's set up with two pages, containing your game's script on the left..

Inform 7 can output Z-Machine story files, the de facto interactive fiction community standard, compatible with the virtual machine used by legendary text adventure developer Infocom in games such as Zork; or Glulx files, which are based on Z-Code but without the older standard's 64kB memory limitation. Unless you're particularly committed to 40 years' worth of backwards compatibility, we recommend opting for Glulx. Glulx interpreters for Raspberry Pi include Gargoyle, ScummVM, and the browser-based Parchment

interpreter. To get started playing Glulx games, open a terminal and type `sudo apt install gargoyle-free`.

If you fancy something a bit more traditional, or if you're an interactive fiction developer working in a language other than English, Inform 6 (**github.com/DavidKinder/Inform6**) is a viable alternative, with a compiler that runs on Raspberry Pi. However, Inform 6 lacks the integrated development environment of its latest successor. The IDE, with its area maps and comprehensive integrated authoring guide, helps to set Inform 7 apart from both its predecessors and many of its rivals.

Install it:
```
sudo apt install flatpak
flatpak remote-add --if-not-exists flathub
https://flathub.org/repo/flathub.flatpakrepo
flatpak install inform
```

Select **app/com.inform7.IDE/aarch64/stable** (option 2 at the time of writing) and type **Y** to install it.
   Approve any other required changes to install dependencies by typing **Y**.
   Inform 7 will be added to the main menu's Programming section.

Notable works:
Anchorhead, Counterfeit Monkey, Hadean Lands.

...You just don't get it, you weren't supposed to be the one arresting me! I DEMAND to see the captain!

Ren'py is the engine underlying numerous very polished commercial visual novels, such as Maxi Molina's The Hayseed Knight

Ren'py

renpy.org

64-bit? Yes (run 32-bit version)
32-bit? Yes

If you've played a modern indie visual novel (VN), it was probably made using Ren'py, a specialist game engine built on top of Python, with a range of almost-natural-language functions designed to do the kind of things you'd expect in a visual novel: load a background scene, show a character, place some building, and display narrative text or dialogue. It's elegant, and the core functions have been kept simple, aided by a superb quickstart guide. This makes it a particularly good choice for artists who wish to start developing their own games.

But you can extend its normal capabilities and features with your own Python scripts, invoked directly from within Ren'py. For example, if you want to add health and combat mechanics to your VN, you'll want to turn to Python. In fact, Python makes Ren'py so flexible that developers are sharing code for minigames ranging from go and solitaire to fishing and lockpicking.

With Ren'Py 8, the entire engine has moved to Python 3, fixing a number of longstanding issues, such as poor handling of accenting and non-Latin characters. Even though Ren'py's official documentation is written to manage your expectations of the engine's ability to run on Raspberry Pi, we found that Ren'py runs in full, with no issues, on Raspberry Pi 4 computers running either 32-bit or 64-bit Linux operating systems, thanks to a dedicated ARM Linux build.

Raspberry Pi 3 can run most Ren'Py games, allowing you to develop for the platform, but may struggle with Ren'py's graphical launcher, depending on your OpenGL settings. This isn't required to develop your games, however, as you'll be spending most of your time using your preferred editor. We use VS Code with the Ren'Py Language extension installed. You can use Ren'Py to develop kinetic visual novels – those that involve no interaction from the reader beyond moving to the next page – as well as VNs with choice-based gameplay and branching dialogues, and even those that incorporate elements of other genres, such as RPG-style combat or the minigames we mentioned above.

One last thing to bear in mind when looking for Ren'py games and sample code is that it's hugely popular for creating adults-only erotic visual novels, and even works that don't feature any sexual content often touch on mature themes. Authors are generally great about providing appropriate warnings about sexual and mature content, but it's worth bearing in mind if you're looking for examples to share with younger developers.

Install it:
Go to **renpy.org** and click on the button to download the version you want – at the time of writing, the latest version is Ren'Py 8.1.0.

Select the 'Download ARM Linux (Raspberry Pi and Chromebook) SDK' link, unzip it, and run `renpy.sh`.

Notable works:
The Hayseed Knight, Loren The Amazon Princess, Long Live The Queen, BBQ DAD.

Build a classic text adventure in Inform

Get started in interactive fiction development

Top Tip 👍

Sentence structure

Although we've organised our code in successive lines for ease of reading, Inform will cheerfully accept paragraphs.

Inform can be used to create very traditional quest-based text adventures, sprawling and intricate works of genre fiction, interactive theatre, and experimental art games. You can add graphics and sound, and a range of extensions (**magpi.cc/intfic**) have been released by the community to add features and functionality. As an introduction to Inform 7's language and capabilities, we're going to create a simple escape room puzzle, with a couple of locations and a handful of objects. In The Shed of Mystery, the player's going to have to work out how to escape a garden shed. You will need a 64-bit installation of Raspberry Pi OS to run the Inform 7 flatpak.

01 Your new adventure

Open Inform 7 and click 'Start a new project'. Click Next to create a new story. Select or create a directory for your project to live in. Enter your game's name and your own, then click Next. Check your project settings, and click Apply.

Inform's main interface opens with an editor on the left and a documentation pane on the right. Tabs on both let you switch between

different content, including extensions, settings, and a map of your previous play-throughs. A play button just above the right-hand pane allows you to play your game. Note the circle arrow next to it: this will replay every command you typed in your previous play-through, allowing you to immediately see the impact of changes to your code.

02 Create a room

We'll start by creating and describing a room. Type the following:

The Shed is a room. 'A cluttered wooden shed, lined with shelves and smelling faintly of creosote.'

The first room that appears in your project is where your player will start the game by default, but you can always override this with a line reading 'The player is in [the room you wish them to start it]' at the beginning.

To help with larger games, you can also group Rooms into Regions, and create Scenes, which can include scripted actions or specific in-game time periods.

03 Your first objects: Supporters

To create an object in Inform 7, you start by naming it. Type:

A workbench is here. The workbench is a supporter. Instead of examining the workbench, say 'The old workbench is almost, but not entirely, free of mess.'

That's created an object called workbench and defined it as a 'supporter', an object which can

▼ Inform's 'skein' records every sequence of actions you've played through, illustrating paths through your game

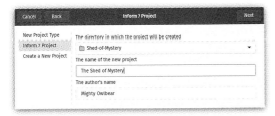

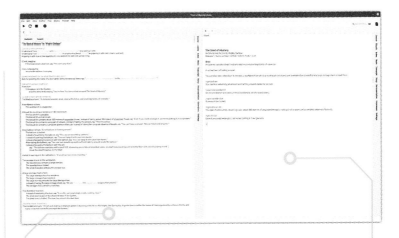

It's a good idea to give all your Inform projects their own directories, just to keep everything tidy

have other objects on top of it. Pay attention to that 'instead of examining' phrase. If we'd just given the workbench a description, as we did the shed, the workbench would be described in full as part of the description of any room that it's in. We only want to show a description when a player LOOKs at the workbench, so our 'instead of' construction provides a specific phrase for that.

04 Inside containers

'Containers' are objects that can contain other objects. Type:

A biscuit tin is a closed container on the workbench. The biscuit tin is openable. The biscuit tin contains a pin.

When you test that code by running the game, you'll be able to OPEN the biscuit tin and TAKE the pin. Now, on a separate line in the editor, type:

A red balloon is here. 'A red balloon is floating around.'

We've given the balloon a description, which, unlike the workbench, will be displayed as part of the description of any room that it's in.

05 Extending Inform's verbs

We've got a pin and a balloon, but Inform doesn't know about popping. Let's change that. Towards the top of our code, we've created a comment – text between square brackets – that reads **[specify new command: pop]**. Below that, we've used an 'understand' statement to describe how our new verb works, and called the 'check' and 'carry out' rulebooks (**magpi.cc/informverb**) to make our 'pop' verb interact correctly with poppable items.

Returning to our red balloon's code, we can now make it poppable and put something inside it. Type:

A small brass key is in the red balloon. 'A small and very shiny brass key.'

Instead of popping the balloon with the pin: say 'The balloon explodes with a loud POP, showering you in bits of shredded rubber. A small brass key drops out onto the floor with a small pinging sound.'; move the small brass key to the Shed.

Perhaps most helpfully, you can keep your code in view while playing through your game. There's also a toolbar button to automatically replay your last play-through, allowing you to change your code on the fly and see how it affects things, among other automated testing features

The Inform IDE's two panes allow you to have your code always visible while you use the second pane to read the extensive built-in documentation and code samples, display compilation errors, and manage settings and extensions

06 Create an exit

The player needs somewhere to escape to and a locked door to prevent them leaving. Type:

The Garden is a room.

It's best to organise rooms and their objects together, so we've put the Garden below everything to do with the Shed. Above it, type:

The shed door is a door. Instead of examining the door, say 'A scruffy door.' The shed door is east of the Shed and west of the Garden. The shed door is locked. The small brass key unlocks the shed door.

07 Winning the game

The player wins when they escape into the garden. At the very top of our code, we've specified a rule that runs every turn to see if they're there:

Every turn:
If the player is in the Garden:
end the story finally saying 'You're free! You have at last escaped The Shed of Mystery!'

Download our extended Shed of Mystery code (**magpi.cc/shedofmystery**) for two more keys to find, extract and use, as well as more items and descriptions. Play with your own shed and see what extra escape room challenges you'd like to add.

Top Tip

Rule of law

Inform 7's syntax is defined by 'rulebooks' such as 'check' and 'carry out'. See **magpi.cc/informverb2** for details.

Top Tip

[magic brackets]

Square brackets are variously used in Inform 7 to define a comment, style text, or as text replacements in text strings or rules.

shed-of-mystery-mini.txt

▶ Language: **Inform 7**

```
001.  "Shed of Mystery Mini" by "Mighty Owlbear"
002.
003.  [check for our win condition]
004.  Every turn:
005.          If the player is in the Garden:
006.                  end the story finally saying "You're free! You have at last escaped
        The Shed of Mystery!"
007.
008.  [specify a new command]
009.  Understand "pop [something]  with [something preferably held]" as popping it with.
010.  Understand "use [something preferably held] to pop/destroy/burst [thing]" as popping it
        with (with nouns reversed).
011.  Popping it with is an action applying to one visible thing and one carried thing.
012.
013.  Check popping:
014.          if the noun is not a balloon, say "You can't pop that."
015.
016.  Carry out popping:
017.          remove the balloon from play
018.
019.  [Create a room and its contents]
020.  The Shed is a room. "A cluttered wooden shed, lined with shelves and smelling faintly
        of creosote."
021.
022.  A workbench is here. The workbench is a supporter. Instead of examining the workbench,
        say "The old workbench is almost, but not entirely, free of mess."
023.
024.  A biscuit tin is a closed container on the workbench. The biscuit tin is openable. The
        biscuit tin contains a pin.
025.
026.  A red balloon is here. "A red balloon is floating around."
027.  A small brass key is in the red balloon. "A small and very shiny brass key."
028.  Instead of popping the balloon with the pin: say " The balloon explodes with a loud
        POP, showering you in bits of shredded rubber. A small brass key drops out onto the
        floor with a small pinging sound."; move the small brass key to the Shed.
029.
030.  The shed door is a door. Instead of examining the door, say "A scruffy door." The shed
        door is east of the Shed and west of the Garden. The shed door is locked. The small
        brass key unlocks the shed door.
031.
032.  [Another room. Freedom! ]
033.  The Garden is a room. "A lush and slightly overgrown garden is bursting with life on
        this bright, late Spring day. A gentle breeze rustles the leaves of trees populated by
        a chorus of birds, and many-coloured insects flit amongst the flowers."
```

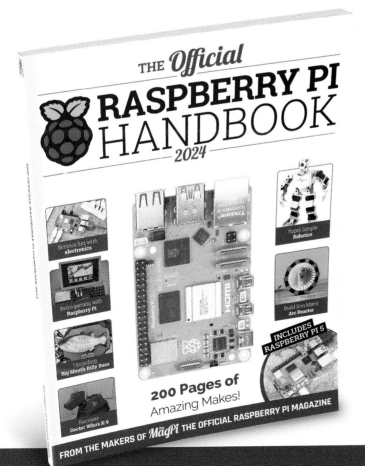

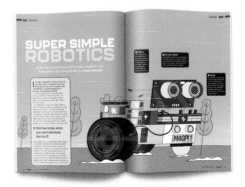

MagPet – code a Python virtual pet

Virtual pets are back! At least here in the magazine, as **Rob Zwetsloot** revives a nineties phenomenon on a Raspberry Pi

MAKER

Rob Zwetsloot

Rob is *The MagPi* Features Editor, and he sometimes fancies himself a game developer when he's not playing games.

magpi.cc

Didya see, didya hear, didya know that virtual pets are still going strong today? Your Tamagotchis, Digimons, and weird alien-themed knock-offs are still around with improved functionality. They're still stuck in their little plastic bodies though, so we thought it was high time we made one from scratch that you can play on your Raspberry Pi desktop. No magical crest required.

Make sure you're on the latest version of Raspberry Pi OS on your Raspberry Pi, and update all the software. Alternatively, you can make this on another computer, as long as you install the Pygame library.

01 Get your art

We've provided code and art on our GitHub at **magpi.cc/magpet**. However, our images are really just placeholders to make sure it all worked.

You'll need at least one image for your pet (ours has a basic two-frame animation cycle), different heart images for happiness, a graphic for when your pet is hungry, a picture of poop, and we also used graphic buttons we made ourselves. Save them in a directory where your Python code is.

Using `pygame.image.load`, you can set the image file to be a variable name, making calling upon it easier in the code.

You'll Need

- Latest Raspberry Pi OS
- Game sprites
- Pygame library (on other systems)

02 Basic parameters

For our version of a virtual pet, we've only imported Pygame (which does include a lot of functions of its own) and the random module for part of the movement cycle.

Our monster friend moves around this enclosed space just living their life

Happiness, hunger, and the state of their living area all affect the monster

FEED   PET   CLEAN

Manage your monster by making sure to clean, feed, and even pet it

The game screen will be 200 pixels by 300 pixels wide in our code. However, if you have large sprites, you may want to increase this.

We've made the background colour the same as a classic Tamagotchi with RGB, and decided to put the pet in the centre of the screen to start. We've also created some global parameters for health and happiness, walk cycle stage, etc., so that the screen updates properly along the way.

03 Move your pet

Our pet is going to move on its own. We'll handle the direction later on but, for now, we have two functions: `pet` and `movement`.

Movement uses cardinal directions translated as numbers (1 = North, 2 = East, etc.) which is selected in the main game code. We move the creature ten pixels in those directions, but also check to see if we're at a boundary so that the pet does not move any further.

Using these co-ordinates, we set the location of the pet in the next frame of the game – we also cycle between the different images that make up the walk cycle.

04 Handle health

What is the goal of this game? To keep our pet happy and healthy. On screen, three hearts are

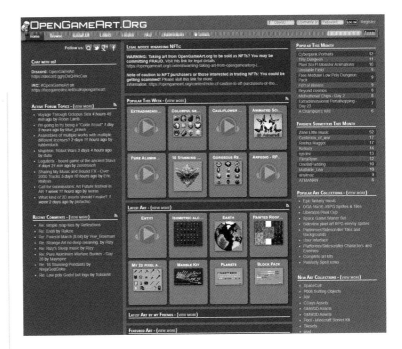

used to show how happy the pet is, with each heart being made up of 20 happiness points behind the scenes. We've created a series of `if` statements to build up the heart graphics, for full, half, and empty hearts.

As the pet relieves itself, it will deposit waste around the screen. The function `poopxy` keeps track of where it's gone, and we're very careful to make sure only one appears at a time. Each time a new poop appears, its co-ordinates get appended to a tuple.

05 Game loop

Each loop of the game, our pet goes through a cycle. It will get hungrier, need to relieve itself more, and also will lose happiness if you're not paying it attention. We've also set it up so happiness is affected by hunger and cleanliness, so make sure to keep your pet fed and clean. When our pet is at a certain level of hunger, a little thought bubble appears to prompt you to feed it.

The `button_pressed` function checks to see where your mouse was when it clicked. This method of tracking pixels is a very basic way to press buttons in Pygame.

06 Main loop

We start the main loop by getting the rest of Pygame set up. This includes a clock to make sure each frame passes at a certain point, creating

For prototyping software that needs graphics, you'll find just about anything you'll need on **OpenGameArt.org** for free

```
1    import pygame
2    import random
3
4    # Game info
5
6    display_width = 200
7    display_height = 300
8
9    bg_colour = (160, 178, 129)
10
11   pet_x = 100
12   pet_y = 100
13
14   # Variables for the pet
15
16   hunger = 2
17   happiness = 20
18   waste = 0
19   wastexy = ()
20   button_press = 0
21   pet_counter = 0
22   walk_cycle = 0
23
24   # Images, including two images for pet idle animation
25
26   pet_1 = pygame.image.load("sprites/pet1.png")
27   pet_2 = pygame.image.load("sprites/pet2.png")
28   full_heart = pygame.image.load("sprites/fullheart.png")
29   half_heart = pygame.image.load("sprites/halfheart.png")
30   empty_heart = pygame.image.load("sprites/emptyheart.png")
31   hungry = pygame.image.load("sprites/hungry.png")
32   poop = pygame.image.load("sprites/poop.png")
33   clean_button = pygame.image.load("sprites/clean.png")
34   feed_button = pygame.image.load("sprites/feed.png")
35   pet_button = pygame.image.load("sprites/pet.png")
36
37   # Game functions
38   # Pet location
39
40   def pet(pet_x,pet_y,game_display):
41       global walk_cycle
42       if walk_cycle == 0:
43           game_display.blit(pet_1, (pet_x,pet_y))
44       else:
45           game_display.blit(pet_2, (pet_x,pet_y))
```

Figure 1 Lines 1 to 35 cover the first two steps. Here, we get everything set up for the rest of the code

Top Tip

What's in a name

Tamagotchi is a portmanteau of tamago/egg and tomodachi/friend in Japanese. They're shaped like eggs after all.

```
47    # Pet movement.
48    # There are checks so it won't move beyond the boundaries
49
50    def movement(move, game_display):
51        global pet_x, pet_y
52        if move == 1:
53            pet_y -= 10
54        if move == 2:
55            pet_x += 10
56        if move == 3:
57            pet_y += 10
58        if move == 4:
59            pet_x -= 10
60        if pet_x < 10:
61            pet_x = 10
62        if pet_x > 190:
63            pet_x = 190
64        if pet_y < 10:
65            pet_y = 10
66        if pet_y > 190:
67            pet_y = 190
68        pet(pet_x, pet_y, game_display)
69
70    # Happiness as displayed by hearts
71    # Each heart is worth 20 points
72
73    def hearts(game_display):
74        global happiness
75        if happiness < 1:
76            game_display.blit(empty_heart, (40,215))
77        if happiness > 0 and happiness < 15:
78            game_display.blit(half_heart, (40,215))
79        if happiness > 14:
80            game_display.blit(full_heart, (40,215))
81        if happiness < 25:
82            game_display.blit(empty_heart, (80,215))
83            game_display.blit(empty_heart, (120,215))
84        if happiness > 24 and happiness < 35 :
85            game_display.blit(half_heart, (80,215))
86        if happiness > 34:
87            game_display.blit(full_heart, (80,215))
88        if happiness < 45:
89            game_display.blit(empty_heart, (120,215))
90        if happiness > 44 and happiness < 55:
91            game_display.blit(half_heart, (120,215))
92        if happiness > 54:
93            game_display.blit(full_heart, (120,215))
94
```

▶ **Figure 2** Step 03 covers the last few lines of **Figure 1**, and also the movement part until line 68. Happiness is shown using the hearts function, explained in Step 04

the game display, and importing all the variables we need.

> ## We decided that it wouldn't just move randomly each frame

Now the full `while` loop starts. The screen is filled with colour, the buttons are placed on the screen, and the mouse co-ordinates are set to 0.

As this part is being called for each frame, it's best to keep as little out of it as you can to make sure your game runs as fast as possible. For this script, the global parameters and unchanging display aspects were placed before the `while` loop for this reason.

Top Tip

Size can matter

The size of your screen and sprites need to match in some way – it's easier to resize images out of Python after all.

07 Handling inputs

In this project, we have two kinds of inputs: a mouse click, which we're using to press the buttons on screen, and also the exit button on the Pygame window.

The `pygame.QUIT` event is fairly simple. If the X on the window is pressed, Pygame will stop. `QUIT`

is the 'event' Pygame receives when you click on the X.

As for the mouse, we wait until the button goes up, hence `MOUSEBUTTONUP` is the event. This means it won't count you having the mouse button depressed as multiple events over several frames if you don't click fast enough. You may have seen variations on this with how buttons work in GPIO Zero.

08 Movement philosophy

Here's where we set the cardinal directions of the pet. We decided that it wouldn't just move randomly each frame, and it would in fact have a 60% chance to carry on in any direction it started. This is where the random module comes in. In action, this means our pet will move more naturally.

The cardinal direction and the `game_display` variable are sent to the `movement` function from

```
95    # Poop location
96
97    def poopxy(waste,pet_x,pet_y,game_display):
98        global wastexy
99        if int(waste) > len(wastexy):
100           wastexy[((int(waste))-1)] = ((pet_x + 5), pet_y)
101           prev_waste = int(waste)
102
103       for i in wastexy:
104           game_display.blit(poop, wastexy[i])
105
106   # Hunger and happiness cycle
107
108   def pet_cycle(pet_x, pet_y, game_display):
109       global hunger, happiness, waste
110       if hunger < 10:
111           hunger += 0.2
112       if hunger > 7:
113           game_display.blit(hungry, ((pet_x + 25),(pet_y - 30)))
114       if happiness > 0:
115           happiness -= 0.05
116           if waste > 3:
117               happiness -= 0.4
118           if hunger > 9:
119               happiness -= 0.2
120       hearts(game_display)
121       if waste < 5:
122           waste += 0.1
123       if int(waste) > 0:
124           poopxy(waste,pet_x,pet_y,game_display)
125
126   # Button locations for press
127
128   def button_pressed(mousex, mousey):
129       global button_press
130       if mousey > 250:
131           return 0
132       else:
133           if mousex >= 0 and mousex <= 60:
134               return 1
135           elif mousex >= 70 and mousex <= 130:
136               return 2
137           elif mousex >= 140 and mousex <= 200:
138               return 3
139           else:
140               return 0
141
```

▲ **Figure 3** Waste is managed as shown in Step 04, and from 106 to the end are the functions that handle your interactions and how your pet lives as shown in Step 05

```
144  def main():
145      pygame.init
146      clock = pygame.time.Clock()
147      game_display = pygame.display.set_mode((display_width, display_height))
148      pygame.display.set_caption("MagPet")
149      move = 0
150      global pet_x, pet_y, happiness, hunger, waste, button_press, pet_counter, walk_cycle
151
152      while True:
153          game_display.fill(bg_colour)
154          game_display.blit(feed_button, (0,250))
155          game_display.blit(pet_button, (70,250))
156          game_display.blit(clean_button, (140,250))
157          mousex = 0
158          mousey = 0
159
160          # Event handler
161          for current_event in pygame.event.get():
162              if current_event.type == pygame.QUIT:
163                  pygame.quit()
164              elif current_event.type == pygame.MOUSEBUTTONUP:
165                  mousex, mousey = current_event.pos
166                  button_press = button_pressed(mousex, mousey)
167
168          # How does the pet move - 1-4 are cardinal directions
169          if move == 0:
170              move = random.randint(0,4)
171
172          if move > 0:
173              if random.randint(1,10) > 4:
174                  movement(move, game_display)
175              else:
176                  move = random.randint(0,4)
177                  movement(move, game_display)
178
179          if button_press != 0:
180              if button_press == 1:
181                  hunger = 0
182                  happiness += 10
183                  button_press = 0
184              if button_press == 2:
185                  if pet_counter > 0:
186                      button_press = 0
187                  else:
188                      happiness += 10
189                      pet_counter = 5
190                      button_press = 0
191              if button_press == 3:
192                  waste = 0
193                  wastexy = ()
194                  button_press = 0
195
196          if pet_counter > 0:
197              pet_counter -= 1
198
199          if happiness < 0 and int(waste) == 5 and hunger > 10:
200              print ("Game Over")
201          else:
202              pet_cycle(pet_x, pet_y, game_display)
203              pygame.display.update()
204
205              if walk_cycle == 0:
206                  walk_cycle = 1
207              else:
208                  walk_cycle = 0
209
210              if pet_counter > 0:
211                  pet_counter -= 1
212
213          clock.tick(2)
214
215
216  if __name__ == '__main__':
217      main()
218
```

Step 3. We pass on `game_display` so that the function can update it properly.

09 Button juggling

There are three buttons on this pet, which means there are four states to keep track of. We've tried to keep it simple and understandable by labelling the buttons 1–3, and no button being 0. One thing we did was to make sure we couldn't just hammer the Pet button to increase happiness. A timer is set so that you have to wait five frames before using it again.

As each frame is rendered separately, this means if you reset hunger and waste, the images for them will not be rendered in the following frame.

> ## We're planning to connect ours to web APIs

10 Game Over…?

How do you lose the game? Well, you need to take care of your pet, so if your pet gets as hungry as it can be (10), fills up its space with waste (5), and loses all happiness (0), it will be game over. On original Tamagotchi devices, they would run away. In our case, the game will just stop and you'll have to start again.

11 Actual play

If your code has made it this far, it's time to update the frame. All the important movement and location functions are called, the cycle is updated, and everything is rendered. As each frame goes by, this gives the illusion of life, and limited time to keep at it.

One important part of the end of the loop is `clock.tick()`. It's used to set a frame rate, with a

Figure 4 This block of code handles Step 06 (144 to 158), Step 07 (160 to 166), Step 08 (168 to 177) and finally Step 09 for the rest

Figure 5 The game is played out through this last bit of code, described in Steps 10 and 11

lower number meaning fewer updates. As virtual pets of old would have very limited movement, we found that 2 simulated a good, slow-moving pet. You can always change it depending on the speed of your hardware.

Finally, the main function is run.

12 Improvements

There's plenty you can do to upgrade this. Better graphics, better buttons, and a tighter adherence to the boundaries as well for starters. You could also add other features like your pet going to sleep for a while, or have it properly run away when it's game over. You could have a status screen with age, and even name it.

We're planning to connect ours to web APIs in the future to see if the internet can properly look after a virtual pet. We don't have high hopes though.

Build a Pico Keypad
Mole Bop game

Create a game of Mole Bop using a 16-button, light-up keypad connected to a Raspberry Pi Pico using a port expander

MAKER

Stewart Watkiss

Also known as Penguin Tutor. Maker and YouTuber that loves all things Raspberry Pi and Pico. Author of *Learn Electronics with Raspberry Pi.*

penguintutor.com

@stewartwatkiss

Important!
Virtual ENV

From Bookworm onwards, packages installed via pip must be installed into a Python Virtual Environment using venv. This has been introduced by the Python community, not Raspberry Pi.

magpi.cc/pipvenv

Understand a way to expand your electronic circuits using I2C. Learn about how to interface with a port expander and see how this can be used for a large number of switches using a switch-array. The first circuit is based around an integrated circuit that you can build on a breadboard. Then, see how a similar technique is used for a 16-button, light-up keypad. Use the keypad to create a fun game of Mole Bop.

01 Using a port expanders

Raspberry Pi Pico has 26 general-purpose input/output pins. For many projects that is more than enough, but sometimes we need more. For a 16-way keypad using a port for each switch and one for each LED, we would need 32 ports.

There are different ways to reduce the number of ports needed. To reduce the number of pins for a keypad, you could use a switch matrix which uses one port for each row and column and scan across them. An example matrix keypad is shown in **Figure 3**. A switch matrix can have a problem with ghosting when multiple buttons are pressed. Sometimes it's easier to use a port expander which increases the number of GPIO ports.

02 Using I2C

I2C is an abbreviation for Inter-Integrated Circuit, which is a protocol for communicating between microcontrollers and integrated circuits. It is a bus technology which allows multiple devices to be connected using only two wires; one for the data and the other for a clock to synchronise the devices. Each device has an address which is used to determine which device to talk to. An example is shown in **Figure 2**.

When using I2C, there needs to be pull-up resistors on the data and clock wires. These can use the internal pull-up resistors available within your Pico; if using devices over a greater distance, external pull-up resistors may be required.

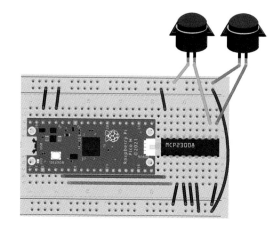

▲ **Figure 1** The MPC20008 is a port expander which can be used on a breadboard as a demonstration of how a port expander can be used

03 Using a MCP23008 port expander

The MCP23008 is an 8-bit port expander which uses I2C to communicate with a Pico or other microcontroller. There are similar devices, such as the MCP23017, which have more ports, but using the smaller MCP23008 means that you can create a circuit on a single breadboard. The example circuit has just two buttons used as a demonstration. This is shown in **Figure 1**.

The address is configurable using pins 3, 4, and 5 on the IC. Connecting all those pins to ground gives an address of 0x20 (hexadecimal value). The

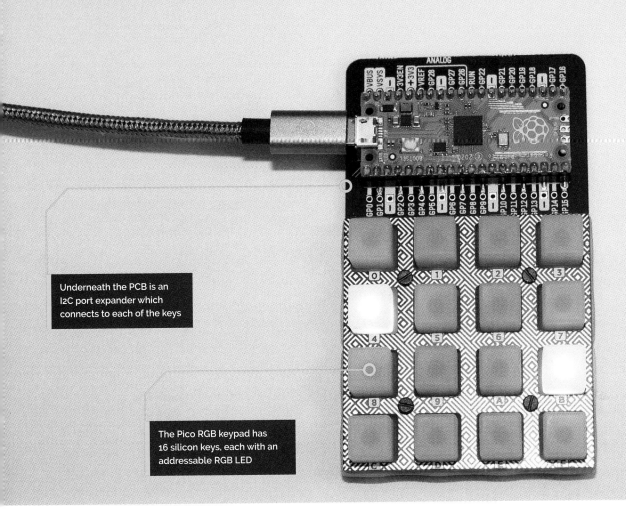

Underneath the PCB is an I2C port expander which connects to each of the keys

The Pico RGB keypad has 16 silicon keys, each with an addressable RGB LED

two wires between the MCP23008 and a Pico are the I2C clock and data signals.

04 Communicating with the port expander

You can use MicroPython to access all the extra ports, almost as easily as if they are local pins. To communicate between a Pico and the port expander, you first need to download the library from **magpi.cc/crankshawnzmp**.

After uploading the library to your Pico, you need to create an instance of I2C and use that to create an instance of the MCP23008 library. The following uses pins 4 and 5 on your Pico and address 0x20 on the port expander:

```
from machine import Pin, I2C
import mcp23008
i2c = I2C(0, scl=Pin(5), sda=Pin(4))
mcp = mcp23008.MCP23008(i2c, 0x20)
```

05 Reading the inputs

The pins can be configured as inputs using the `setPinDir` command. This takes the port number as the first argument and the value 1 for the second argument to indicate

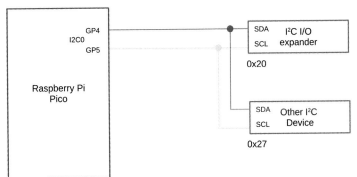

that it is used as an input (0 would be used for output).

```
mcp.setPinDir(0,1)
```

You can also enable pull-ups for the buttons using `setPullupOn`.

```
mcp.setPullupOn(0)
```

Once it is configured, you can query the status using `readPin`.

```
mcp.readPin(0)
```

This gives a 1 when the button is pressed, otherwise a 0.

▲ **Figure 2** I2C is a bus protocol allowing multiple devices to be connected using two GPIO ports on Raspberry Pi Pico

You'll Need

> 2 × button switches **magpi.cc/sanwa**

> MCP23008 port expander **magpi.cc/ MCP23008**

> Pico RGB Keypad Base **magpi.cc/ picokeypad**

Top Tip 👍

Other I2C devices

There are many other devices which use I2C, including various sensors and output devices such as LCD screens.

06 Pico RGB keypad base

Using a port expander, you could connect that to buttons and LEDs to create a keypad. Alternatively, you could buy a Pico RGB keypad base which has already wired a keypad on a custom PCB with a 4×4 silicone keypad. The keypad uses a TCA9555 port expander on the PCB (**Figure 4**). This is like the MCP23008, but has 16 input/output ports and is a surface-mount package (better for PCBs, but not so good for breadboards). The chip even uses the same I2C address of 0x20. The PCB also includes an APA102 addressable LED for each key. These are like NeoPixels, but use a protocol based on SPI instead of the single data connection that NeoPixels use.

07 Preparing the keypad base

The base needs a little assembly for the buttons, then you plug a Pico using header pins. Instead of installing the drivers separately, Pimoroni provides a version of MicroPython with all the drivers pre-installed. Download the Pico file from **magpi.cc/pimoronipicogit**. Transfer the UF2 file to your Pico after pressing the BOOTSEL button during startup. The keypad and LEDs can be accessed by importing picokeypad and creating an instance of that class.

```
import picokeypad
keypad = picokeypad.PicoKeypad()
```

Figure 3 An alternative is to use a matrix keypad. These have the buttons arranged in a grid of rows and columns, which reduces the number of pins needed

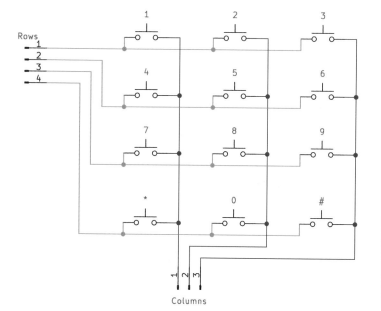

08 Reading the keypad

The keypad is read by using the `get_num_pads` method. This returns a value between 0 and 65535 depending upon which buttons are pressed. There are 16 keys, but we've only used the first eight to simplify the explanation. **Figure 5** shows an overview of how the binary number system works in relation to the keypad numbers.

Inside the microcontroller, the status of the keys is represented by a binary digit for each key. If the key is pressed, then it has a value of 1 and, if it is not pressed, then it has a value of 0. The rightmost digit is key 0 which, if pressed, has a decimal value 1. If button 1 is pressed, that has a value of 2. That goes up to 128 if button 7 is pressed.

09 Shifting the values

To determine which button is pressed, you can read each of the bits in turn and see which are set to 1.

Starting at the right, look at the rightmost digit which is a 0, which means that button 0 is not pressed. Shift the numbers to the right and the rightmost digit becomes a 1, which means that button 1 is pressed. In Python, the shift is done using >> (two greater-than characters) followed by the number of bits to shift.

To perform the comparison, the bitwise AND function is performed. If you compare using '& 0x01' that will ignore all but the rightmost bit.

10 Creating a game

To turn this into a game, a random button is chosen which is then lit up. The player must try and press that button before the timer runs out. If the player presses the button before `lit_duration` is passed, then they score a point. If they don't, then the button turns red and the player loses

Keypad number	7	6	5	4	3	2	1	0
	128	64	32	16	8	4	2	1
Binary digit	1	0	0	0	0	0	1	0

Figure 5 An example of the binary code showing keypad numbers 1 and 7 pressed. The numbers above the binary digits are the decimal equivalent for each key

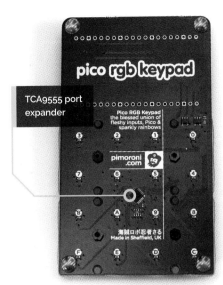

TCA9555 port expander

Figure 4 The TCA9555 port expander is the small square integrated circuit mounted on the rear of the keypad PCB

a life. After all lives are lost, the grid lights up green to show the final score, with each button representing four points.

11 Adding a challenge

To make the game more challenging, each time a point is scored, the time for the next button is reduced by 30 milliseconds. Whenever a life is lost, it is increased by 500ms to give a reasonable chance to score more points. This makes it possible to get a reasonable score, with the difficulty increasing during the game.

Raspberry Pi Pico does not have a real-time clock, so to work out the time that has elapsed, `time.ticks_ms` is used. This measures time in milliseconds. The method `ticks_diff` is used to test whether the `lit_duration` time has been exceeded.

12 Further development

The game has some areas for improvement. Firstly, it will only run once and then you need to restart it from Thonny. Another `while` loop can be added to keep the game running, along with resetting the score and other variables.

Whilst it is possible to get a range of scores, many players will score around the same amount. This is due to the way that the complexity increases linearly. An alternative is to increase the difficulty using levels instead.

Another improvement would be to have multiple buttons light at the same time. See version two in the GitHub repository for an improved version of the game.

mole-bop.py

 Language: **MicroPython**

```python
001.  import time
002.  import random
003.  import picokeypad
004.
005.  keypad = picokeypad.PicoKeypad()
006.  keypad.set_brightness(0.5)
007.  NUM_KEYS = keypad.get_num_pads()
008.
009.  lit_button = random.randint(0, NUM_KEYS - 1)
010.  lit_time = time.ticks_ms()
011.  lit_duration = 1000
012.
013.  score = 0
014.  lives = 3
015.
016.  def all_off():
017.      for i in range (0, NUM_KEYS):
018.          keypad.illuminate(i, 0,0,0)
019.
020.  while True:
021.      # Check to see if the time has expired
022.      now_time = time.ticks_ms()
023.      if (time.ticks_diff (now_time, lit_time) > lit_duration):
024.          # light it red for missed
025.          keypad.illuminate(lit_button, 255, 0, 0)
026.          keypad.update()
027.          time.sleep (1)
028.          lives -= 1
029.          if (lives <=0):
030.              break
031.          # make the next one a little easier by adding 400ms
032.          lit_duration += 500
033.          # choose new button
034.          lit_button = random.randint(0, NUM_KEYS - 1)
035.          lit_time = time.ticks_ms()
036.
037.      # Turn LEDs off and the lit one on
038.      all_off()
039.      keypad.illuminate(lit_button, 255, 255, 255)
040.      keypad.update()
041.
042.      # scan keys to see if the lit_button key is pressed
043.      button_states = keypad.get_button_states()
044.      for i in range (0, NUM_KEYS):
045.          if i == lit_button and button_states & 0x01 > 0:
046.              score += 1
047.              lit_duration -= 30
048.              lit_button = random.randint(0, NUM_KEYS - 1)
049.              lit_time = time.ticks_ms()
050.          button_states = button_states >> 1
051.
052.  print ("Score {}".format(score))
053.  all_off()
054.  if (score > 64):
055.      score = 64
056.  for i in range (0, score / 4):
057.      keypad.illuminate(i, 0, 255, 0)
058.  keypad.update()
```

Make a Pico LCD true or false quiz game

An LCD display can display messages from Raspberry Pi Pico. In this project, the display is used as part of a quiz game

MAKER

Stewart Watkiss

Also known as Penguin Tutor. Maker and YouTuber that loves all things Raspberry Pi and Pico. Author of *Learn Electronics with Raspberry Pi*.

penguintutor.com
twitter.com/
stewartwatkiss

I n this tutorial, we will create an interactive true or false quiz game using a Raspberry Pi Pico and an LCD display. Whilst doing so, you will learn about some of the pitfalls when connecting to devices running at different voltages. You'll discover ways to increase voltage output using a simple buffer, and how to make a bidirectional level shifter. The level shifter is then used to convert between 3.3V for the GPIO ports on a Pico to 5V used by the LCD display. The game is programmed in Python, with a text file for the questions. The game can be installed inside an enclosure for a complete game.

You'll Need

> LCD display with PCF8574T
> **magpi.cc/ihaospacelcd**

> I2C safe level-converter
> **magpi.cc/BSS138**

> 3 × 16 mm button switches
> **magpi.cc/16mmbutton**

01 LCD character display

This project is based around an LCD display. Our display has 16 characters across two lines and is often referenced as a '1602'. These usually contain an HD44780, or equivalent, driver chip that displays the appropriate pixels that make up the characters.

One downside of the display is that the driver chip needs at least six data connections. This uses up GPIO ports, as well as needing lots of wires to the LCD display. A common solution is to have a 'backpack' fitted to the rear of the LCD display using a port expander. The example used here is a PCF8574T 8-bit port expander.

02 Designed for 5 V

The port expanders are available on a PCB backpack pre-soldered onto the back of the LCD PCB. This saves you from having to create your own circuit, but it does come with an issue. These circuits are normally designed for 5V, whereas a Pico uses 3.3V for the GPIO ports.

Connecting a 5V signal to a Pico GPIO port could cause permanent damage to the latter, so this tutorial looks at some of the possible solutions to interfacing between devices designed for different voltages.

03 Move pull-up to 3.3 V

If the 5V device did not have a pull-up resistor, the I2C bus could work with pull-ups to the 3.3V supply instead. This is shown in **Figure 2**. The crossed-out resistors are the pull-ups inside the LCD I2C backpack and the two pull-up resistors on the left are connected to the 3.3V output on a Pico. Unfortunately, this involves de-soldering surface-mount devices, which can be difficult.

04 Unidirectional level shifter

A simple form of level shifter can be used when controlling 5V devices from a 3.3V microcontroller or computer. This is often used for controlling NeoPixels from a Pico or a Raspberry Pi. In its simplest form, this is a MOSFET with two resistors (as shown in **Figure 3**, overleaf). The gate resistor RG (typically 470 Ω) reduces the in-rush current, and RL is a pull-up resistor (typically 2.2 kΩ to 10k Ω). With no input, the pull-up resistor sets the output high. When a 3.3V input is provided, the MOSFET turns on pulling the output low. This results in an inverted signal.

The code can be configured to invert the output, or you could add an additional MOSFET to invert it a second time. A two-stage, non-inverting buffer is shown in **Figure 4**.

05 Bidirectional level shifter

The LCD is controlled from your Pico, so you may expect the signal would only need to go in one direction. However, due to the use of I2C protocol, signals need to pass in both directions. We need a bidirectional level shifter. These can be made using individual MOSFETS, but using a premade level shifter from Adafruit or SparkFun is more convenient. An example is the Adafruit bidirectional level shifter, which has four level shifters on a convenient PCB. This is shown in **Figure 5**.

The level shifter has just one MOSFET for each channel. This is in an unusual configuration. The circuit can be thought of as two sides, with the left side being for the low voltage and the right for the higher voltage. The MOSFET joins the two together. The schematic diagram is shown in **Figure 6**.

06 How the level shifter works

If both the low-voltage and high-voltage signals are high, then the MOSFET is off and the signal is high at both sides. If the low-voltage signal (left) drops low, then the MOSFET is in the forward direction and the voltage at the gate will turn the MOSFET on. This will provide a path to ground and so the high-voltage signal (right) will be pulled low. If the high-voltage signal (right) goes low, due to an internal characteristic of the MOSFET a small current is able to flow in the reverse direction. As this happens, the voltage of the source pin dips, causing the MOSFET to turn on. This pulls the voltage down on the low-voltage signal as well.

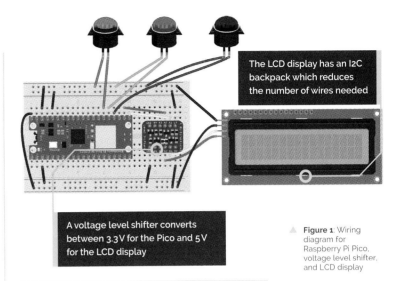

The LCD display has an I2C backpack which reduces the number of wires needed

A voltage level shifter converts between 3.3 V for the Pico and 5 V for the LCD display

Figure 1: Wiring diagram for Raspberry Pi Pico, voltage level shifter, and LCD display

07 LCD circuit

The level shifter can be inserted onto the breadboard and connected between your Pico and LCD display. Then it's just a case of adding three buttons for Start, True, and False. These are shown in **Figure 1**.

The top power rail is used for 3.3V taken from your Pico's 3.3V output, and the bottom power rail is 5V taken from the VBUS supply from the USB port.

The buttons used are 16 mm push-to-make switches, similar to arcade buttons, but smaller. You can use other push-to-make switches if you prefer.

08 Download the LCD library

The libraries that support the LCD display with backpack are available from GitHub (**magpi.cc/rpipicoi2clcd**). Upload the files

Important!
Virtual ENV

From Bookworm onwards, packages installed via pip must be installed into a Python Virtual Environment using venv. This has been introduced by the Python community, not Raspberry Pi.

magpi.cc/pipvenv

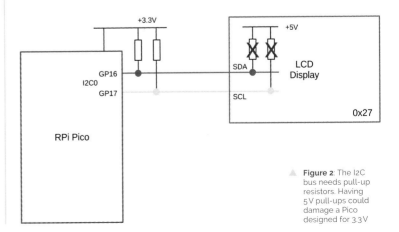

Figure 2: The I2C bus needs pull-up resistors. Having 5 V pull-ups could damage a Pico designed for 3.3 V

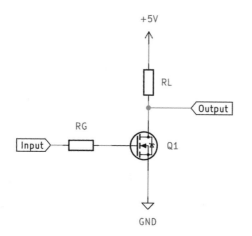

+5V

RL

Output

RG

Input

Q1

GND

▶ **Figure 3**: A simple MOSFET level shifter. The output is the opposite of the input but higher voltage

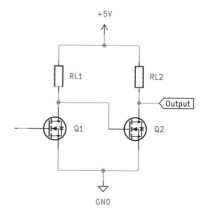

+5V

RL1   RL2

Output

Q1   Q2

GND

▲ **Figure 4**: A second MOSFET can be used to create a non-inverting buffer

lcd_api.py and **pico_i2c_lcd.py** to your Pico. You can see a demo using **pico_i2c_lcd_test.py**. This can be useful for checking your wiring is correct, but you will need to change the pins used for SDA (GPIO 16) and SCL (GPIO 17).

09 Coding the game

The game code (**quizgame.py**, overleaf). starts by setting up the three `button` objects, along with `i2c` and `lcd`. It then reads the file **quizfile.txt**, which contains the questions.

Then it enters a loop which ensures that the game can be played over again.

Within the first few lines of the loop, you can see that it first clears the display, puts a string which starts on the top line, moves to the start of the second line, and then puts another string to that line.

10 Handling button presses

The button presses are handled by having a `while` loop which runs until an appropriate button is pressed. In the case of the Start button, it just looks for that one button, but when waiting for a true and false, it needs to check both the

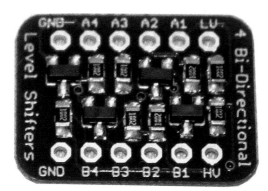

▶ **Figure 5**: A bidirectional level shifter is available on small PCBs. These can have headers fitted to be used on a breadboard

`true_button` and `false_button` to see if either is pressed. It keeps track of the score and then displays the score at the end, pausing for five seconds before restarting the game.

11 The quiz file

The questions are stored in the file **quizfile.txt**. This has one line per question. Each line should have three entries separated

❝ The code to create the game is included ❞

by a semicolon. The first entry is the top line to display, the second is the second line, and the final entry is a letter T or F to indicate whether the correct answer is `True` or `False`.

The file is opened using the `with` statement. Using with means that the file will be automatically closed after the program has finished reading in the entries. The `readlines` method is used to read all the entries into a list.

To separate the text to display from the answers, the `split` method is used. You may notice that it also uses the `strip` method to ignore any whitespaces, such as spaces before the newline character.

The quiz file is created separately and must be uploaded to Pico.

12 Improving the game

The game can be placed in an enclosure to make a complete game. You could start with a standard enclosure and cut holes for the display and buttons, or if you have a 3D printer you can

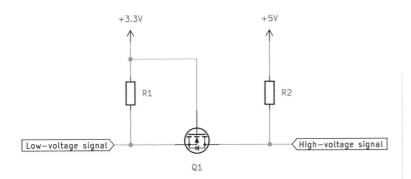

Figure 6: The bidirectional level shifter uses two MOSFETs. This works well for signals with pull-up resistors like I2C

Figure 7: The game can be placed inside an enclosure. If you don't have a 3D printer, you can use a generic case and cut appropriate holes

download an example from the Penguin Tutor website (**magpi.cc/trueorfalse**). One improvement would be to add some error checking. Without error checking, if there is an invalid entry in the quiz file, the program may crash.

Another possible improvement would be to provide a way to add multiple quizzes rather than just limiting them to a single quiz. ◼

quizgame.py

> Language: **MicroPython**

```
001.    import utime
002.    from machine import Pin, I2C
003.    from lcd_api import LcdApi
004.    from pico_i2c_lcd import I2cLcd
005.
006.    I2C_ADDR     = 0x27
007.    I2C_NUM_ROWS = 4
008.    I2C_NUM_COLS = 16
009.
010.    start_button = Pin(20, Pin.IN, Pin.PULL_UP)
011.    true_button  = Pin(19, Pin.IN, Pin.PULL_UP)
012.    false_button = Pin(18, Pin.IN, Pin.PULL_UP)
013.
014.    i2c = I2C(0, sda=machine.Pin(16),
        scl=machine.Pin(17), freq=400000)
015.    lcd = I2cLcd(i2c, I2C_ADDR, I2C_NUM_ROWS,
        I2C_NUM_COLS)
016.
017.    lcd.hide_cursor()
018.
019.    # Read questions into a list
020.    with open("quizfile.txt", 'r') as file:
021.        questions = file.readlines()
022.
023.    while True:
024.        lcd.clear()
025.        lcd.putstr("True or False")
026.        lcd.move_to(0,1)
027.        lcd.putstr("Press Start ...")
028.
029.        while True:
030.            if (start_button.value() == 0):
031.                break
032.        # Quiz start
033.        score = 0
034.        for question in questions:
035.            lcd.clear()
036.            # strip off any whitespace
037.            # then split the entries into line1, 2 and answer
038.            (text, text2, answer) =
        question.strip().split(";", 3)
039.            lcd.putstr(text)
040.            lcd.move_to(0,1)
041.            lcd.putstr(text2)
042.            while True:
043.                if (true_button.value() == 0):
044.                    if (answer == "T"):
045.                        score += 1
046.                    break
047.                if (false_button.value() == 0):
048.                    if (answer == "F"):
049.                        score += 1
050.                    break
051.            lcd.clear()
052.            lcd.putstr("Game over")
053.            lcd.move_to(0,1)
054.            lcd.putstr("Score {} of {}".format(
        score, len(questions)))
055.            utime.sleep(5)
```

Build a Pong-style game with Raspberry Pi

Create a version of the classic pong game using a Raspberry Pi.
This project creates rotating paddles using potentiometers and
an SPI analogue-to-digital converter

MAKER

**Stewart
Watkiss**

Also known as
Penguin Tutor.
Maker and YouTuber
that loves all things
Raspberry Pi and
Pico. Author of
*Learn Electronics
with Raspberry Pi.*

penguintutor.com

twitter.com/
stewartwatkiss

This game of bat and ball is based around a
Raspberry Pi with potentiometers for the
paddle controllers, to mimic the classic
game from the 1970s. As a Raspberry Pi does
not include any analogue inputs, an analogue-
to-digital converter is used. This uses the Serial
Peripheral Interface (SPI) which provides a way
for a computer or microcontroller to communicate
with integrated circuits and other devices. Python
and Pygame Zero are used to create a graphical
display with the Pong-style paddles and ball.

the values as a digital value for a Raspberry Pi or
microcontroller. The MCP3008 has eight inputs
which can be used individually, or as pairs, to
provide voltage comparison. It converts the
analogue value into an 8-bit number between 0
and 1023.

There are Python drivers available, but they are
no longer supported. Instead, a simplified version
of the library is provided in the GitHub repository:
magpi.cc/rpiponggit. Save the file **mcp3008.py**
into the same directory as your own code.

You'll Need

> MCP3008 ADC
magpi.cc/mcp3008

> 2 × 10K potentiometers
magpi.cc/10kpot

> Mini breadboards
**magpi.cc/
minibreadboardwhite**

01 Raspberry Pi

This project features a Raspberry Pi
computer, which has an operating system and
more processing power than Raspberry Pi Pico.
It can connect direct to a screen using HDMI
allowing us to create a graphical game. One thing
it doesn't have are analogue inputs, which are
included in a Raspberry Pi Pico.

This tutorial will show how an external
analogue-to-digital converter (ADC) can be added
to Raspberry Pi and used to control a game that
makes use of Pygame Zero.

02 MCP3008 ADC

The MCP3008 is an integrated circuit
which can sample analogue inputs and provide

03 SPI

Raspberry Pi uses Serial Peripheral
Interface (SPI) to communicate with the
MCP3008. This is a protocol designed primarily
for communicating with embedded systems.
It is similar to I2C, but has an additional data
line and requires a chip select connection for
each peripheral device, instead of using device
address numbers.

Figure 2 (overleaf) shows the connections
between the controller and peripheral devices.
The two data lines are labelled as SDO (serial
data out) and SDI (serial data in). You may also
see these referred to as MOSI and MISO. The CS
lines are for chip select. These determine which
peripheral the controller is communicating with.
The line above CS indicates that the signal is
inverted (it is active when set to a low signal).

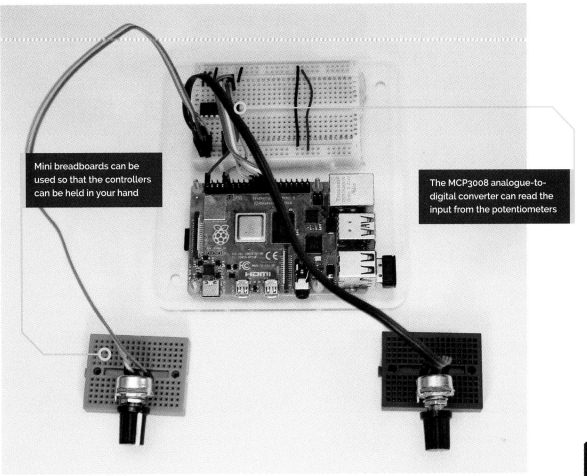

Mini breadboards can be used so that the controllers can be held in your hand

The MCP3008 analogue-to-digital converter can read the input from the potentiometers

04 Wiring up the MCP3008

The circuit diagram is shown in **Figure 1**. The data out on your Raspberry Pi is pin 19 (GPIO 10 / MOSI), which is connected to the MCP3008 pin 11 (digital in). Data in on your Raspberry Pi is pin 21 (GPIO 9 / MISO), which is connected to the MCP3008 pin 12 (digital out). The clock signal is on pin 23 of your Raspberry Pi (GPIO 11), which goes to the clock input on pin 13 of the MCP3008. Finally, chip enable 0 is used on your Raspberry Pi from pin 24 (GPIO 8) to pin 10 on the MCP3008 (chip select). See **Figure 3** for the MCP3008 pinout. You should also enable SPI using the Raspberry Pi Configuration tool (**Figure 4**).

05 Wiring up the potentiometers

The power supply for the potentiometers and the MCP3008 comes from Raspberry Pi's 3.3V power supply. For convenience, this is connected to the bottom and top power rails on the breadboard. Looking from the top of the potentiometer, the left pin connects to ground and the right pin to the 3.3V supply. The wipers then

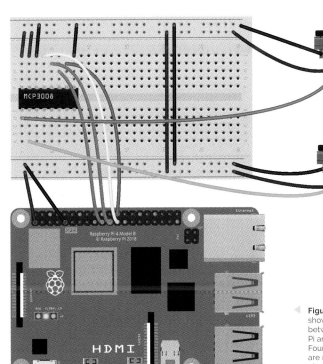

Figure 1: Diagram showing the wiring between a Raspberry Pi and the MCP3008. Four connections are required from physical pin numbers 19, 21, 23, and 24

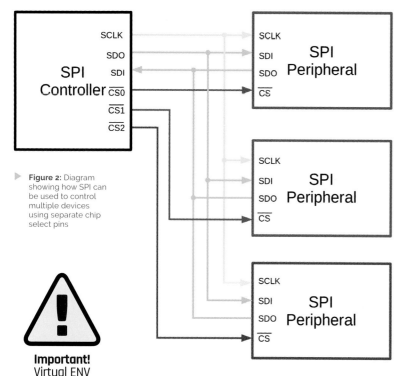

Figure 2: Diagram showing how SPI can be used to control multiple devices using separate chip select pins

⚠️

Important!
Virtual ENV

From Bookworm onwards, packages installed via pip must be installed into a Python Virtual Environment using venv. This has been introduced by the Python community, not Raspberry Pi.

magpi.cc/pipvenv

Top Tip 👍

SPI terminology

The terminology used is based on the latest guidelines from the Open Source Hardware Association. For more details see **magpi.cc/ spisignalnames**.

▶ **Figure 3:** The pinout of the MCP3008 analogue-to-digital converter. The input channels can be used individually or in pairs

connect to pins 1 and 2 of the MCP3008, which are the first two analogue inputs. This should provide 0 V when the potentiometer is turned to the left. The wires could be soldered directly to the potentiometers, or jumper leads could be used to mini-breadboards.

06 Pygame Zero

The game needs a way to draw the graphics on the screen. One of the easiest ways of getting started creating graphics is with Pygame Zero. Using Pygame Zero, you can draw on the screen with minimal setup code.

Creating the game window is as simple as specifying the window dimensions using:

```
WIDTH = 1280
HEIGHT = 720
```

To run from the command-line, use `pgzrun` followed by the file name. You can also run the game using Thonny by selecting Pygame Zero mode under the Run menu.

07 Draw and update functions

To simplify code, Pygame Zero uses two functions, `draw` and `update`, each of which runs approximately 60 times per second. The `draw` function is used for drawing the screen. In this case: it clears the screen to remove the previous frame. It checks to see if the game has finished

using `game_state` and if so displays a message to the screen using `screen.draw.text`. Otherwise, the game is in progress and it calls the functions to draw the two paddles and the ball.

The `update` method handles all the rest of the logic, including getting the position for the paddles from the ADC, updating the position of the ball, and detecting whether the ball hits the paddles or goes off the screen.

08 Get the paddle position

The position of the paddles is based on the reading from the MCP3008. This is read using:

```
adc0 = mcp.read_adc(0)
```

The value returned is between 0 and 1023, which needs to be converted to a value representing the position on a 720-pixel-high screen. This is done by dividing the value by a height constant of 1.5 and subtracting it from the screen height (as the y-axis is numbered from the top downwards). The size of the paddle is also subtracted to allow part of the paddle to go off the end of the screen.

This is an approximation, but allows the paddle to cover the full height of the screen without completely disappearing.

Note that the code refers to player 0 and player 1, but when displayed to the user they are shown as player 1 and player 2 respectively. This is to provide consistency with the Python lists which count from 0.

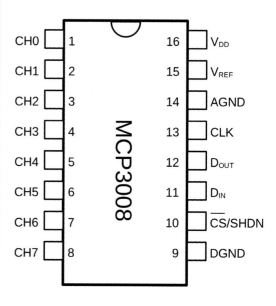

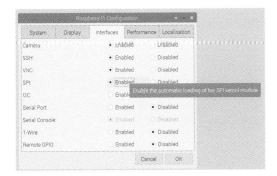

Figure 4: SPI needs to be enabled through Raspberry Pi OS's configuration tool under the Preferences menu

09 Creating the paddle rectangle

The paddle on the screen is drawn using a `Rect` object. This is a Pygame class which describes a rectangle. As well as being used for drawing the paddle, there are built-in methods that help with collision detection, which will be explained later.

The `Rect` object takes the x, y coordinates of the top left corner as the first argument and then width and height as a second. Player 1 has a `Rect` defined as below:

```
Rect((40, player1_pos), (paddle_width, paddle_
height))
```

The value of 40 is the distance from the left of the screen and `player1_pos` is the vertical position. The width and height of the paddle are constant values but could be changed in a future version to change the difficulty. The `Rect` objects are stored in a global list called `paddle_rect`.

10 Drawing shapes in Pygame Zero

The paddles are drawn using the `draw_paddle` function which in turn calls the `filled_rect` draw method for the relevant player.

The ball is drawn using the `draw_ball` method, which draws the ball as a filled circle based on the ball's x and y coordinates.

In both these examples, the `color` variable is set to white to mimic the classic game, but they could be changed to provide different-coloured paddles and ball.

11 Game play

To turn this into a game, the ball needs to move around and detect when it's hit something. This is handled within the `update` function.

The ball has a list for the velocity which determines the speed of the ball in the x and y directions; this is added to the position of the ball.

The code then checks to see if the ball has hit the left or right side, in which case it sets the state

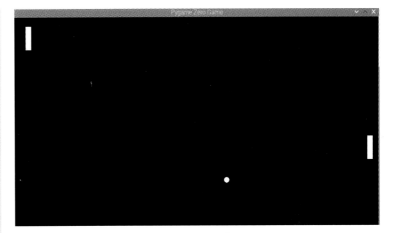

The basic game with white paddles and a ball

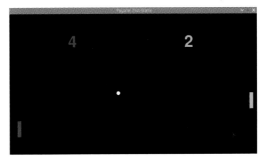

A second version of the game is included in the GitHub repository showing the score and coloured paddles

to `gameover` and sets the winner. It then checks if the ball has hit a paddle, which is done using the `Rect collidepoint` method. If the ball has hit a paddle, the `deflect_ball` function is used to change the x direction and, depending upon where on the paddle, the speed in the y direction. It also checks to see if the ball has hit the top or bottom, in which case the y direction is changed.

Finally, the speed is increased each turn to increase the difficulty as the game is played.

12 Future improvements

At the moment, the game ends after a single point is scored. To improve the game, this can be expanded to handle scores and the score displayed on the screen. This is included in an updated version called **pong2.py** included in the GitHub repository.

You may wish to add the ability to run the game a second time, or even add different games. This may need a button for each controller so that players can select when to start the game. You could even create other games using the paddle controllers, such as a Breakout clone.

Top Tip 👍

Pygame Zero

Pygame Zero is designed for education and makes creating your own games as simple as possible. For more information see **magpi.cc/ pygamezero**.

pong.py

> Language: **Python Pygame Zero**

```
001.  # Pong using MCP3008 and two potentiometers
002.  # Run through Pygame Zero (pgzrun)
003.  import time
004.  from mcp3008 import MCP3008
005.
006.  WIDTH = 1280
007.  HEIGHT = 720
008.
009.  mcp = MCP3008()
010.
011.  WHITE = (255,255,255)
012.
013.  player0_pos = mcp.read_adc(0)
014.  player1_pos = mcp.read_adc(1)
015.
016.  #starting positions
017.  ball_x = 640
018.  ball_y = 360
019.  ball_speed = 5 # default 5
020.  ball_velocity = [1 * ball_speed, 0.5 *
      ball_speed]
021.  ball_radius = 10
022.
023.  paddle_height = 80
024.  paddle_width = 20
025.
026.  # Rectangle representing paddles
027.  paddle_rect = [(0,0),(0,0)]
028.
029.  # Ratio screen (720) to adc (1024)
030.  height_constant = 1.5
031.
032.  game_state = "play"
033.  game_winner = 0
034.
035.  def draw():
036.      screen.clear()
037.      if (game_state == "gameover"):
038.          # To make more user friendly change
          player 0 to 1, and 1 to 2
039.          screen.draw.text("Player {} wins!".
      format(game_winner+1), (100, 50), fontsize=60)
040.      else:
041.          draw_paddle(screen, 0)
042.          draw_paddle(screen, 1)
043.          draw_ball(screen, ball_x, ball_y,
      ball_radius)
044.
045.  def update():
046.      global paddle_rect, player0_pos,
      player1_pos, ball_x, ball_y, ball_velocity,
      game_state, game_winner
047.      adc0 = mcp.read_adc(0)
048.      adc1 = mcp.read_adc(1)
049.      player0_pos = HEIGHT - int(
      adc0 / height_constant) - paddle_height
050.      player1_pos = HEIGHT - int(
      adc1 / height_constant) - paddle_height
051.      paddle_rect[0] = Rect((40, player0_pos),
      (paddle_width, paddle_height))
052.      paddle_rect[1] = Rect((WIDTH-40,
      player1_pos), (paddle_width, paddle_height))
053.      # Update ball position
054.      ball_x += int(ball_velocity[0])
055.      ball_y += int(ball_velocity[1])
056.
057.      # Hits wall = game over
058.      if (ball_x - ball_radius <= 0):
059.          # player 1 wins if passes left
060.          game_winner = 1
061.          game_state = "gameover"
062.      elif (ball_x + ball_radius >= WIDTH):
063.          # player 0 wins if passes right
064.          game_winner = 0
065.          game_state = "gameover"
066.          # Collides with paddle - checks centre
      of the ball
067.      elif (paddle_rect[0].collidepoint (ball_x,
      ball_y)):
068.          deflect_ball (0)
069.      elif (paddle_rect[1].collidepoint (ball_x,
      ball_y)):
070.          deflect_ball (1)
071.
072.      # Hit top or bottom (bounce)
073.      if (ball_y + ball_radius >= HEIGHT or
      ball_y - ball_radius <= 0):
074.          ball_velocity[1] = ball_velocity[1] *
      -1
075.
076.      # Increase speed of ball - quick increase
077.      # Max speed 20
078.      if ball_velocity[0] < 20 and
      ball_velocity[0] > -20:
079.          ball_velocity[0] =
      ball_velocity[0] * 1.001
080.          ball_velocity[1] =
      ball_velocity[1] * 1.001
081.
082.  # player left = 0, right = 1
083.  def draw_paddle(screen, player, color=WHITE):
084.      if player == 0:
```

```
085.            screen.draw.filled_rect(paddle_rect[0],
       color)
086.        else:
087.            screen.draw.filled_rect(paddle_rect[1],
       color)
088.
089.  def draw_ball(screen, ball_x, ball_y,
       ball_radius=10, ball_color=WHITE):
090.        screen.draw.filled_circle ((ball_x,ball_y),
       ball_radius, ball_color)
091.
092.  def deflect_ball (player):
093.        global ball_velocity, ball_x, ball_y
094.        # reverse horizontal direction
095.        ball_velocity[0] = ball_velocity[0] * -1
096.        # also move the ball 2 pixels away to
       prevent it getting stuck
097.        if player == 0 :
098.            ball_x += 2
099.        else :
100.            ball_x -= 2
101.        # if moving down
102.        if ball_velocity[1] > 0:
103.            if ball_y < paddle_rect[player].y +
       (paddle_height / 3):
104.                ball_velocity[1] = ball_velocity[1]
       * 0.5
105.            elif ball_y > paddle_rect[player].y + 2
       * (paddle_height / 3):
106.                ball_velocity[1] = ball_velocity[1]
       * 1.5
107.            # If central part of paddle then no
       change
108.        else:
109.            if ball_y < paddle_rect[player].y +
       (paddle_height / 3):
110.                ball_velocity[1] = ball_velocity[1]
       * 1.5
111.            elif ball_y > paddle_rect[player].y + 2
       * (paddle_height / 3):
112.                ball_velocity[1] = ball_velocity[1]
       * 0.5
113.            # If central part of paddle then no
       change
114.
115.
116.
117.
118.
119.
120.
```

BUILD RETRO MACHINES

CONSOLES, COMPUTERS AND ARCADE MACHINES

Emulate a BBC Micro on Raspberry Pi 400

With an optimised version of the B-Em emulator, Raspberry Pi 400 can take us back to the heyday of the British micro that inspired it

MAKER

K.G. Orphanides

KG is a writer, developer and software preservationist.

magpi.cc/ hauntedowlbear

You'll Need

> Pico SDK
> **magpi.cc/picosdk**

> Pico Extras
> **magpi.cc/ picoextras**

> Kilograham's B-Em fork
> **magpi.cc/bem**

F rom schools to bedrooms, Acorn Computers' BBC Micro helped to kick-start a revolution in software and hardware development. Its influence is imprinted on products ranging from the ubiquitous Arm processors to modern gaming blockbusters like Elite Dangerous.

But emulating the BBC on modest hardware has traditionally been a chore. There are several excellent BBC emulators, and you can run most of them on Raspberry Pi, but only at the cost of major performance issues.

That's no longer the case, thanks to Raspberry Pi Pico SDK architect Graham 'kilograham' Sanderson who's forked an optimised version of the B-Em emulator that'll run smoothly on both the Raspberry Pi Pico microcontroller and Raspberry Pi SBCs.

> Make sure that the Pico SDK, Pico Extras, and B-Em source code folders are all in the same directory before you compile

01 Check your graphics settings

When building this B-Em fork for Raspberry Pi, you need to make sure that the compositor is disabled and that you're not using the legacy video driver. If you're running a fresh Raspberry Pi OS Bullseye install on Raspberry Pi 400, neither of these should be the case, but it's worth checking.

To do so, open a Terminal window and type:

```
sudo raspi-config
```

Select '6 Advanced', then 'A3 Compositor'. Make sure this isn't enabled. If your hardware and OS versions support the legacy video driver, you'll also find this in the advanced menu. Similarly, make sure it's disabled. Reboot.

02 Gather your dependencies

At the Terminal type:

```
sudo apt install build-essential cmake
libdrm-dev libx11-xcb-dev libxcb-dri3-dev
libepoxy-dev ruby libasound2-dev
```

Although we're building B-Em for Raspberry Pi, we'll still need the Pico SDK. The Pico SDK, Pico Extras, and B-Em folders all need to be in the same directory. We'll make a directory called **Pico** in our home root to put everything in, then put all the software we need in it.

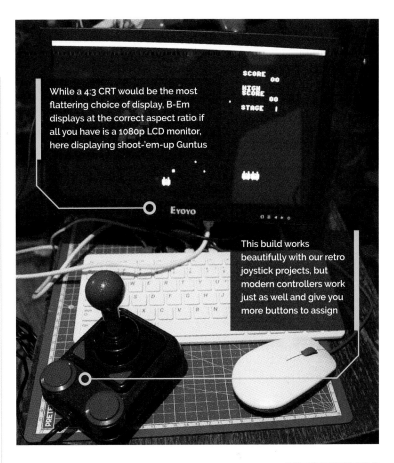

You should substitute the wget URL in the next step with the current release version of the Pico SDK (**magpi.cc/picosdk**).

While a 4:3 CRT would be the most flattering choice of display, B-Em displays at the correct aspect ratio if all you have is a 1080p LCD monitor, here displaying shoot-'em-up Guntus

This build works beautifully with our retro joystick projects, but modern controllers work just as well and give you more buttons to assign

03 Build B-Em

Enter the following Terminal commands to download and make the software:

```
cd
mkdir Pico
wget https://github.com/raspberrypi/pico-sdk/archive/refs/tags/1.4.0.tar.gz
tar -xzf 1.4.0.tar.gz
git clone https://github.com/raspberrypi/pico-extras.git
git clone https://github.com/kilograham/b-em.git
cd b-em
mkdir pi_build
cd pi_build
cmake -DPICO_SDK_PATH=path/to/pico-sdk -DPI_BUILD=1 -DPICO_PLATFORM=host -DDRM_PRIME=1 -DX_GUI=1 ..
make -j4
```

Following these instructions, the path to Pico SDK will be something like: **~/Pico/pico-sdk-1.4.0**.

04 Is this thing on?

B-Em emulators for the BBC Micro (xbeeb) and Master (xmaster) will be built two directories below the **pi-build** directory we made earlier, in **src/pico**. Assuming you still have the Terminal open from the previous step:

```
cd src/pico
./xbeeb
```

You should find yourself looking at an emulation of a 32K BBC Micro running Acorn's 1770 DFS (disk filing system).

05 Would you like to play a game?

Unlike other BBC emulators, including the fully loaded versions of B-Em, there's no graphical interface to browse and load disk image files. Fortunately, B-Em for Pico loads a copy of Eben Upton's port of puzzle game 2048 by default. Not

all the keys and characters on Raspberry Pi 400 map tidily to those of a BBC Micro – see **Figure 1** overleaf. In B-Em, type:

```
*.
```

…to list the files on the virtual disk, then:

```
*EXEC !BOOT
```

…to run the program.

06 Key combinations and commands

In this version of B-Em, **F11** opens the menu, navigable using the mouse or arrow keys. Press **ENTER** to select an option and, if it flashes

◀ This version of B-Em has a stripped-down **F11** menu but it includes critical sound and video options, plus embedded disc image handling

Top Tip 👍

Build for Pico!

After this tutorial, you'll also be set to build B-Em for Raspberry Pi Pico. You'll need a VGA board to handle graphics output.

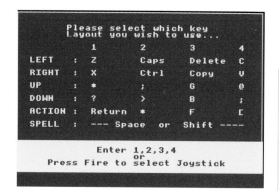

There weren't many consistent standards for game or cursor control layouts in the early 1980s, leading to some creative approaches, demonstrated here by Citadel

after selection, to confirm it. Press **ESC** to exit the menu or opt out of a flashing selection. The pared-down menu gives you sound and graphics settings, access to embedded disks, and a hard reset. To exit B-EM, close the window or press **ALT**+**F4** to terminate the program.

We've already seen that `*.` lists the files on a disk and `EXEC` executes a program. You can find a helpful lookup sheet for other commands and for BBC BASIC in general at **magpi.cc/bbcbasic**. An HTML version of the BBC Micro User Guide exists at **magpi.cc/bbcuserguide**.

07 Loading disk images

Although there's no disk image browser, you can launch B-Em to run a specific disc image. The .SSD and .DSD formats are supported, these are the most commonly used to distribute modern BBC games.

Download Infuto's Guntus shoot-'em-up from **magpi.cc/guntus** (direct download) and save it to your **Downloads** folder.

To load it, the command with full paths should be:

```
~/Pico/b-em/pi_build/src/pico/xbeep
~/Downloads/Disc172-Guntus.ssd
```

Figure 1 BBC Micro keyboard map for the UK Raspberry Pi keyboard. BBC keys are marked in red

We'll make some shortcuts to tidy that up later. In B-Em list the files by typing `*.`, then load Guntus with:

```
CHAIN"GUNT-IN"
```

We could also have used `*EXEC !BOOT`, as a boot file is present.

08 Embedding disks

You can also embed disk images at build time. By default, B-Em boots with two embedded disks: 2048 and the iconic BBC Welcome disk. You can switch between them in the **F11** menu, by going to Disc 0, using the left and right arrow keys to navigate to the option you want, then pressing **ENTER**.

In our **~/Pico/b-em/src/pico/discs/** directory, you'll find a file called **beeb_discs.txt**, alongside several disk images. Add new embedded disks by inserting a line at the bottom formatted thus:

```
Your Label For The Disk = /path/to/disk/
image.dsd
```

Alternatively, you can create a file in the same directory, called **beeb_discs_user.txt**, with entries formatted in the same way. This will automatically supersede **beeb_discs.txt**. Labels can have spaces; paths and file names cannot.

Once you've added your list of files to embed, repeat the final `make -j4` command in the **pi_build** directory from Step 3.

09 Sticks of joy

Joystick support isn't built into this version of B-Em, but you can use mapping software to work around this. AntiMicroX can map any recognised joystick, from a Sony DualShock 4 to the Pi Pico-powered DB9 joystick adapter we made in issue 125, or our older GPIO joystick adapter (**magpi.cc/db9joystick**). At the Terminal:

```
sudo apt install antimicro
antimicrox
```

A popular control schema for games used **Z** to move left, **X** to move right, **:** for up, and **/** for down, with **SPACE** and **RETURN** configured as jump and/or action keys. If you only have one joystick button, assign it to whichever you need most often. AntiMicroX supports multiple control profiles, so it's easy to switch between the most common.

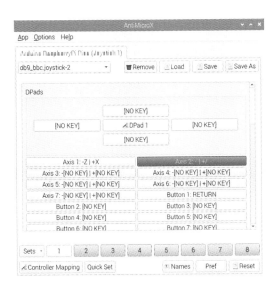

Mapping a controller with AntiMicroX is a convenient way of both adding retro joystick support to this version of B-Em, and replacing old-school keyboard layouts with something more familiar

b-em.desktop

> Language: **DESKTOP**

```
001.  [Desktop Entry]
002.  Encoding=UTF-8
003.  Version=1.0
004.  Type=Application
005.  Terminal=false
006.  Name=B-Em
007.  Comment=BBC Micro emulator
008.  Exec=~/Pico/b-em/pi_build/src/pico/xbeeb
009.  Icon=~/Pico/b-em/icon/b-em.png
```

10 A smoother startup

Let's add some symlinks and shortcuts Open a Terminal and type:

```
sudo ln -s ~/Pico/b-em/pi_build/src/pico/
xbeeb /usr/bin/local
```

You can now launch xbeeb by name at the Terminal, no path required. You'll still need to provide paths to disc files you wish to launch, though.

We'll also make a desktop launcher. Create a text file called **b-em.desktop** on the desktop, or add it to Raspberry Pi OS's main menu by placing it in **~/.local/share/applications**. See the code listing for the full launcher file. If you want to launch from the desktop, embed the disks you use regularly, as there's no file browser. To ensure that .desktop files are executed by default, in the file manager go to Edit > Preferences and tick 'don't ask options on launch executable file'.

11 Getting new software

There's a vibrant homebrew scene for the BBC Micro. The non-commercial section (**magpi.cc/bbcnoncomm**) of the Complete BBC Micro Games Archive is home to 29 games released in 2022 alone, plus hundreds of older freeware and public domain titles ranging from the 1980s to the present. New releases are usually announced on the forums of mainline B-Em and BeebEm maintainer group Stardot: **magpi.cc/bbcstardot**.

> ❞ There's a vibrant homebrew scene for the BBC Micro ❞

12 Learn to program like it's 1985

Many of Usborne Books' classic 1980s guides to computer programming are now available for free at **magpi.cc/usborne1980s**. BBC Micro projects feature heavily.

There are even modern publications that review and analyse software for 8-bit platforms including the BBC Micro, such as *The Classic Adventurer* (**classicadventurer.co.uk**), while historians and enthusiasts have archived contemporary magazines and fanzines at sites including the Educational Software Archive for the BBC Micro (**magpi.cc/bbceduarchive**), 8BS (**magpi.cc/8bsmag**), and The Internet Archive (**magpi.cc/microuserarchive**).

Top Tip

Now I am the Master

B-Em includes a BBC Master emulator, which you'll find in the same directory, with an executable file named **xmaster**.

▼ Modern BBC Micro games like Brian Tierney's Blue Wizard are able to use impressive graphical techniques not readily available to BBC devs in the early 1980s

Beepy: Play games on a palmtop computer

Who needs Candy Crush when you can play
Colossal Cave Adventure with a full keyboard?

MAKER

K.G. Orphanides

KG is a developer and writer who has been waiting over a decade to play text adventures at the bus stop without having to fiddle with an on-screen keyboard.

twoot.space/@ owlbear

You'll Need

> SQFMI Beepy kit or equivalent
> **beepy.sqfmi.com**

> 3D printed case (optional)
> **magpi.cc/slimflat**

Beepy is a handheld computer that can be set up with the basics, including a word processor, spreadsheet, encrypted instant messaging and more. That's great, but handheld devices can be idle distractions and sources of stimulation, too. While it's not going to replace either the Game Boy or the Steam Deck, Beepy is extraordinarily well suited to keyboard-driven activities, from parser text games to extremely online social media posting. We'll look at books, games and other distractions for Beepy. Remember that you can turn **ESWD** into arrow keys by pressing the mod key (the touch button in the middle of the bar below Beepy's screen), or activate the touchpad itself by double-tapping it.

01 Read a book

Terminal ebook reader epy (**magpi.cc/ epy**) supports the EPUB, EPUB3, FB2, MOBI, AZW and AZW3 formats. Last month, we used Python's pip package management system within a venv to install a Python 3 application, matrix-commander, and keep it discrete from other Python applications on our system. This time, we'll use the pipx installer, which automates a lot of this for you. Via an SSH session, type:

```
$ pipx install epy-reader
```

Make a directory for your ebooks, and grab a public domain epub to read – we'll use a terminal browser to download a book from Project Gutenberg, which doesn't allow wget or curl access to individual books.

```
$ mkdir Ebooks
$ cd Ebooks
$ w3m https://www.gutenberg.org/ebooks/11.
epub.noimages
```

Once the file has been downloaded, press **Q** to quit, then we'll rename it:

```
$ mv pg11.epub AlicesAdventuresInWonderland-
NoImages.epub
$ epy AlicesAdventuresInWonderland-NoImages.
epub
```

Navigate the book with **EWSD**, press **Q** to quit.

02 Toot on Mastodon

If you thought building a handheld PC might help you ditch your social media habit, you're out of luck, but interacting with it in the terminal might at least curb your worst excesses. You'll need to create an account on a Mastodon server before you begin this step. If you don't have one, there's a list at **joinmastodon.org/ servers**. SSH to Beepy and then type:

```
$ sudo apt install toot
$ toot login
```

Say No when asked if you'd like to open the authentication link in your default browser.

Toot can connect to any Mastodon account. In this case, that's K.G.'s self-hosted server.

Posts are ordered by date and time – as you scroll down the left-hand column, the contents of each post are displayed on the right

03 Authenticate and toot

Instead, copy the supplied login URL from your SSH terminal, open it in a browser, make sure you're logged into the correct account, and click Authorize.

Copy the authorisation code, return to your SSH session to Beepy and paste in the code.

You should see a message telling you that you've successfully logged in.

Now, grab Beepy, open a new Tmux session if you want to keep the app open in the background, and type:

```
$ toot tui
```

Press the trackpad button to switch into ESWD cursor navigation or use vi-style HJKOL, and enjoy your very, very Usenet-like fediverse browsing experience. Quit with **SHIFT**+**Q**. Read the full docs at **magpi.cc/tui**.

04 RSS never died

Whether you use a synchronised service or just want to subscribe to a clutch of feeds locally, Newsboat (**newsboat.org**) is an outstanding RSS reader. We'll install and run Newsboat once to create its config directory, which by default is `.newsboat` in your home.

```
sudo apt install newsboat
newsboat
nano ~/.newsboat/urls
```

Now add the URLs of any RSS feeds you wish to subscribe to. For example, we added **https://newsboat.org/news.atom** and **https://www.raspberrypi.org/blog/feed/**. Press **CTRL**+**X** to save and exit and pick up Beepy.

05 All aboard the newsboat

At the terminal, type:

```
$ newsboat
```

You'll notice that your feeds are listed, but that there's nothing in them. Press **SHIFT**+**R** to load all feeds, or highlight an individual feed and press **R** to reload only that feed. You can

▲ While not every roguelike lends itself to Beepy's 400×240 display, Nethack and its relations work a treat

navigate by pressing the number of each feed or by using **N** (next) and **P** (previous). You can scroll through messages with the trackpad or by putting Beepy into ESWD directional key mode. Enter opens a feed or a message, while **Q** moves up a level, and quits Newsboat if you're at the top level/

Alternatively, you can create a `~/.newsboat/` config file to sync to an external RSS feed reader service, such as a self-deployed Nextcloud News installation, The Old Reader, or FreshRSS. See **magpi.cc/newsboatclient** to find the appropriate settings for your service.

▼ With newsboat, you can subscribe to all your favourite blogs and periodicals, download them when you have a Wi-Fi connection, and read them at your leisure

06 Get on IRC

IRC (Internet Relay Chat) is the Ur online messaging protocol, pre-dating and inspiring the likes of XMPP, Slack and Discord. It remains active, with networks including Libera.Chat, QuakeNet, Undernet and EFnet. Find more at **magpi.cc/ircnetworks**.

```
$ sudo apt install irssi
$ irssi
$ /network add LiberaChat
$ /server add -network LiberaChat -tls -tls_
verify irc.libera.chat 6697
$ /connect LiberaChat
$ /join #raspberrypi
```

This unofficial Raspberry Pi enthusiasts' channel tends to be quiet, but works for this example. Many servers, including Libera Chat, require you to register an account and nickname for full access to all channels, particularly busier general #chat channels. Libera Chat provides detailed instructions at **magpi.cc/liberachatreg**.

07 Terminal games

There are plenty of good (and less good, but nonetheless interesting) games that you can play in your terminal, but not all of them adapt well to Beepy's 400×240 mono display, while others are too fiddly to play with the ESWD directional key layout. Thus, our recommendations don't include the terminal port of trading simulation strategy classic *Taipan*, or roguelike RPG classic *Angband*. It's a good idea to create a games directory and ensure that the ncurses development libraries are installed. Where relevant, we'll also symlink games to `/usr/local/games`, so you don't have to type the full path to play them.

```
$ sudo apt install libncurses5-dev
$ mkdir Games
```

08 Tiny Moon Runner

The only game so far in Ben Busby's ncurses arcade (**magpi.cc/ncursesarcade**), it's like that T-rex cactus jumping game you can play when Chrome isn't working, and it's great. Press **SPACE** to jump over rocks as an ASCII art landscape scrolls by, or be smashed to your doom.

Freesweep is an ncurses-based Minesweeper clone. All the default settings - except colour - are Beepy compatible. You can navigate with ESWD in arrow key mode, or vi-style with HJLK. Press **F** to place a flag and **SPACE** to clear a square. Quit with **Q**.

Internet Relay Chat is long past its heyday, but remains an excellent way to chat, without the vast system overhead of Slack or Discord

10 Dungeon delving

To play *Nethack*, enter:

```
$ sudo apt install nethack-console
```

While most roguelikes in the Debian games bundle require far too much screen real estate, this console version of *Nethack* works well, and you can control your movement with the vi keys, HJKL, although the limited number of keys means it's easiest to use # commands for tasks such as looting chests. You'll find the command reference at **magpi.cc/nethackcommands**.

▼ Terminal ebook reader epy allows you to carry a library with you on Beepy. You'll need to use DRM-free ebooks with it, though

```
$ cd Games
$ git clone https://github.com/benbusby/
ncurses-arcade.git
$ cd ncurses-arcade
make
./ncurses-arcade
$ sudo ln -s /home/YOUR_USER_NAME/Games/
ncurses-arcade/ncurses-arcade /usr/local/
games
```

You'll now be able to run it from anywhere by typing:

```
$ ncurses-arcade
```

09 Debian's terminal game pack

You can bulk install Debian's collection of terminal games via the games-console meta-package, but the majority don't work properly on the hardware available, most often due to Beepy's petite display size. We've tested all of them, and can recommend a number that actually work, from RPGs to procedural plant generators. Enter:

```
$ sudo apt install freesweep
```

▲ Support for Z-Machine and Glulx files opens up a world of interactive fiction. Seen here, Andrew Plotkin's commercial Glulx release, Hadean Lands

```
$ sudo apt install slashem
```

Nethack variant *Slashem* (which stands for Super Lotsa Added Stuff Hack – Extended Magic) is also fully functional. *Slashem* gets extra player character races, including vampires, hobbits and drow, as well as new monsters, spells and more. However, it's generally regarded as more difficult than the original, largely due to game balance issues with the new levels and monsters.

11 Classic spelunking

If you fancy a text adventure, enter:

```
$ sudo apt install colossal-cave-adventure
```

This is a faithful port of Will Crowther's early text adventure, inspired by caver Crowther's exploration of Mammoth Cave in Kentucky, and it's where familiar phrases like "you are in a little maze of twisty passages, all alike" come from. This port is pretty faithful to the 1976 version, in which you can earn a maximum of 350 points by exploring and collecting items.

```
$ sudo apt install open-adventure
```

This is another port of *Colossal Cave Adventure*, this time expanded upon by Don Woods in 1977. It has quality life features including not yelling at you in capital letters all the time. This is probably the one you want to play, as it pays less attention to the accurate reproduction of the cave complex and more to being a game. Launch it by typing **advent**. There are 430 points to be won in this iteration.

Both versions are pretty tough, so you may want some hints. Try **magpi.cc/colcavehints**.

12 The world of interactive fiction

The text adventures of the 1970s gave birth to a rich culture of interactive fiction. Beepy's keyboard makes it the ideal device for portable parser gaming, because you can easily type in commands such as GET LAMP. In that spirit, we're going to install Glulxe, an IF engine virtual

> ❝ Beepy's keyboard makes it the ideal device for portable parser gaming ❞

machine that lets you play parser games written in the Glulx format supported by Inform 6.30 and above, enter:

```
$ sudo apt install glulxe fizmo-ncursesw
```

We're going to install a couple of classics: the original free version of Michael Gentry's Lovecraftian horror game *Anchorhead* (**magpi. cc/anchorhead**), later developed for commercial release, and Emily Short's *Counterfeit Monkey*, a word-bending puzzle adventure. You can find more free games at the Interactive Fiction Database (**ifdb.org**).

```
$ cd Games
$ wget https://ifarchive.org/if-archive/
games/zcode/anchor.z8
$ fizmo-ncurses anchor.z8
$ wget https://github.com/i7/counterfeit-
monkey/releases/download/r11.1/
CounterfeitMonkey-11.gblorb
$ glulxe CounterfeitMonkey-11.gblorb
```

Top Tip

Adventure online

Install tintin++ to access multiplayer dungeons and other online text games. Learn more at **magpi.cc/ tintinmud**.

Make a SpecDeck:
Digital tape loader for the ZX Spectrum

Fed up with 'R Tape Loading Error' errors on your classic Speccy?
Load ZX Spectrum software with ease using Raspberry Pi and Pirate Audio

PJ Evans

PJ has owned a ZX Spectrum since 1982. He's pictured here with Richard Attwasser and Dr Steven Vickers, the hardware and software designers of the ZX Spectrum.

twitter.com/ mrpjevans

Forty years ago, Sir Clive Sinclair brought home computing to the masses with his affordable ZX Spectrum. It was soon the centrepiece of living rooms across the UK as young gamers and coders battled with parents for control of the television. As wonderful a machine as it still is, the tape loading system for games and other software was fraught with errors. Wonky cassettes and unreliable playback hardware caused no end of frustration. Now we can use a Raspberry Pi Zero to emulate a tape onto an original ZX Spectrum with a reliability you could only have dreamt of in 1982.

You'll Need

> Pirate Audio Headphone Amp HAT
magpi.cc/ pirateaudiohead

> Headphone amplifier
magpi.cc/ compactstereoamp

> 2 × 2.5˝ audio cables

> 3D-printed case (optional)
magpi.cc/ pirateaudiocase

01 Raspberry Pi, Assemble!

Our first challenge with this project is volume. ZX Spectrums like it loud and the audio output from Model A and B Raspberry Pi computers just isn't enough for these old computers to hear. So we start by adding a Pimoroni Pirate Audio HAT, specifically the Headphone Amp version, which will get us some of the way to an acceptable output level. It also means we can use the more portable Raspberry Pi Zero W for this project. Start by carefully adding the HAT to Raspberry Pi Zero (it'll need GPIO headers, so a WH variant is perfect). If you're not planning on using a case, we recommend adding some pillars for stability.

02 Prepare your SD card

It's software time now. We don't need a full-blown OS, so Raspberry Pi OS Lite is perfect for our needs. If you're building this project headless (with no keyboard or monitor attached), then we recommend the new advanced features in Raspberry Pi Imager (**magpi.cc/imager**). Run Imager and select Raspberry Pi OS (Other) then Raspberry Pi OS Lite as the image and your SD card as the storage. Now click the cog to set the hostname, enable SSH, create an account, and set wireless LAN credentials. Now you can write your image and, on boot, your Raspberry Pi will connect to the network and be ready to go.

03 Enable and test Pirate Audio

By now you should be able to boot your Raspberry Pi and gain access via SSH. Don't worry if you've connected a monitor and keyboard: the instructions are exactly the same. Now we need to enable Pirate Audio. Log in and enter the following command:

```
sudo nano /boot/config.txt
```

Then, at the bottom of the file, add these lines:

```
dtoverlay=hifiberry-dac
gpio=25=op,dh
```

The SpecDeck converts digital versions of ZX Spectrum files to audio and plays them back just like a cassette player

This vintage ZX Spectrum loads games by listening to tones from a cassette player and converting them into digital information

```
dtparam=audio=off
```

Save (**CTRL+X**) then reboot (`sudo reboot`). After reboot, connect the audio out to headphones (careful of the volume!) and run this command:

```
speaker-test -c 2
```

Do you hear static-like white noise? If so, you're good to go. Press **CTRL+C** to stop the racket.

04 Will it work? It depends on dependencies

The lines we added to **config.txt** in the previous step did two things: enable audio output to the DAC (digital-to-analogue converter) on the HAT and disable any other audio output. However, to get control over the screen, we have to install libraries that will help us write and display data. These are known as dependencies. Start by making sure you have updated everything already installed to the latest version:

```
sudo apt -y update && sudo apt -y upgrade
```

Once complete, you can install all the libraries you need:

```
sudo apt install git libsdl2-mixer-2.0-0
python3-rpi.gpio python3-spidev python3-pip
python3-pil python3-numpy libatlas-base-dev
libportaudio2
```

Finally, run `sudo raspi-config` and under Interfacing Options, enable SPI and I2C.

05 Turn it up to 11

What you do next depends on what kind of hardware you are intending to use with this project. If it is an emulated ZX Spectrum such as FUSE or Spectaculator, or a modern recreation such as Harlequin, you can skip this. However, if you want to load to an original ZX Spectrum, including later variants such as the 128+, you'll need further amplification. What will do the trick

Top Tip

Keep it down

Please be careful when testing playback, ZX Spectrums need it loud, so please don't listen directly to playback on headphones! If you need to test hold them away from you, you'll be able to hear!

The Pirate Audio HAT is a DAC (digital-to-analogue converter) combined with a small headphone amp

is a standard battery-powered headphone amp. You need to connect its input to the Pirate Audio HAT and the output to the ZX Spectrum. It will provide the final boost needed for reliable loading.

06 Soundcheck

We've now got enough of the project working to test loading on a real ZX Spectrum (or whatever you are using) but we need something to load. We provided a test program. To download it, enter this at the command line:

```
wget https://github.com/mrpjevans/specdeck/
raw/main/raspberrypi.wav
```

Once downloaded, hook up the output from your amplifier to the EAR socket on the ZX Spectrum, turn it up to max, run **LOAD ""** on the ZX Spectrum and enter:

```
aplay raspberrypi.wav
```

If all is well, the program will start to load!

specdeck.service

DOWNLOAD THE FULL CODE:

⊙ **magpi.cc/ specdeckservicepy**

> Language: **Service**

```
001.  [Unit]
002.  Description=specdeck
003.
004.  [Service]
005.  ExecStart=/usr/bin/python3 /home/pi/specdeck/specdeck.py
006.  Restart=on-failure
007.
008.  [Install]
009.  WantedBy=multi-user.target
```

07 Playing with Python

For our project, we want to be able to select from a number of games, and also handle conversion from the compact TZX format directly to WAV. We're going to create a Python app that will do this for us, as well as allow playback control and display cover art on the Pirate Audio's tiny display. To demonstrate how you can control playback of audio in Python, take a look at the **specdeck_1.py** code listing. We need the amazing Pygame module and a few other libraries to help us first:

```
sudo pip3 install pygame keyboard st7789
tzxtools
```

Now, if you enter the code into a file called **playback.py** in the same directory as the WAV file, you can start playback with:

```
sudo python3 playback.py
```

08 Get the full code

The final project code is a bit long for these pages, so we've provided an easy way to download everything and get running. From your home directory, run the following:

```
git clone https://github.com/mrpjevans/
specdeck.git
```

This will download everything you need. The main file is called **specdeck.py**, but we've also included several iterations of the project (**specdeck_1.py** to **specdeck_6.py**) to show how the program evolved from the original code listing to the finished version. This method of coding is much more satisfying than trying to build everything all at once. You'll find everything in the **specdeck** directory.

09 Add some files

In the **specdeck** directory, you will find three further directories; **tzx** takes the standard ZX Spectrum TZX tape image format. When played for the first time it will be converted into a WAV file and placed in the **wav** directory, so we don't have to convert it again. If a JPG file is found with the same name in the **image** directory, it is loaded and displayed on the screen (it will be automatically

This inexpensive headphone amplifier can be combined with the Pirate Audio HAT to produce a sound loud enough for a ZX Spectrum to load reliably

resized if needed). You can add as many files as you like here from legitimate sites such as World of Spectrum (**worldofspectrum.org**). We've provided our test program so you can try things out immediately.

specdeck_1.py

> Language: **Python 3**

```
001.  from time import sleep
002.  import pygame.mixer
003.  pygame.mixer.init()
004.  pygame.mixer.music.set_volume(1)
005.  pygame.mixer.music.load("./raspberrypi.wav")
006.  pygame.mixer.music.play()
007.  while True:
008.      sleep(0.1)
```

10 Add some files

From the command line, run this:

```
cd ~/specdeck
sudo python3 specdeck.py
```

After a few seconds, 'SpecDeck!' will be displayed on the screen and then Raspberry Pi's logo. This means all is working. Pressing button A on the HAT will start conversion of the TZX to WAV. This takes a little time on the Raspberry Pi Zero, but the resulting WAV will be kept so it will be instant next time. You can pause playback by pressing button A again and rewind by pressing B. On the right-hand side, X and Y will scroll through all the TZXs available. Press **CTRL+C** to stop the program.

11 Spectrum service

We don't want to have to log in via SSH every time we want to use our SpecDeck, so let's start everything up on boot by installing the program as a system-level service.

First, create a new file:

```
sudo nano /usr/lib/systemd/specdeck.service
```

Now add the text in the **specdeck.service** listing and use **CTRL+X** to save and exit. Now we have our service file, we can enable it to run on boot:

```
sudo systemctl enable /usr/lib/systemd/
specdeck.service
```

Here we can see a healthy loading tone: the equal bars in the border show that we have the levels just right

Finally, test it with a reboot:

```
sudo reboot
```

On startup, you should see the 'SpecDeck!' announcement and then our test file. You can shut down your Raspberry Pi safely by pressing and holding button B for five seconds.

12 Add some files

Our SpecDeck is a little exposed and there's lots of delicate stuff that needs protecting. Fortunately, there are a lot of cases that people have designed for the Pirate Audio HAT that can be 3D printed. Our favourite, by Yasuhiro Wabiko, can be downloaded here: **magpi.cc/pirateaudiocase**. This can add that all-important protection and make the SpecDeck truly portable. Other improvements could be to add a battery to make it truly portable, or add support for other computers such as the ZX81 or Commodore 64. As ever, it's over to you.

Top Tip

Legal ROMs

Looking for content for your SpecDeck? The ZX Spectrum homebrew scene is as busy as ever. See **magpi.cc/legalroms**.

Build a DB9 to USB joystick converter

Connect retro joysticks to a modern PC over USB using Raspberry Pi Pico

MAKER

K.G. Orphanides

KG is a writer, developer, and software preservationist. They own one of those 1990s joysticks shaped like Giger's Alien and can confirm that it's not very comfy to game with.

magpi.cc/ hauntedowlbear

You'll Need

▸ Arduino IDE 2.0.2 or above
arduino.cc/en/ software

▸ PicoGamepad library
magpi.cc/ picogamepad

▸ DB9 joystick INO file
magpi.cc/ picoconverter

▸ A DB9 joystick

▸ Serial port (male)

We'll be using Raspberry Pi Pico to convert input from a 9-pin DB-9 joystick, via male D-sub port, into USB input recognised by any modern PC. We refer to these 9-pin D-sub connectors as 'DB-9' – because it is the most widely used term – but they are more correctly called DE-9 connectors, which you'll also find used. It's the same physical connection as an RS232 serial port and a D-sub port, which makes for a long list of terms used interchangeably.

We'll be programming Raspberry Pi Pico using the cross-platform Arduino IDE and Arduino's Mbed OS for RP2040 boards. This gives us access to Arduino's mature and very capable HID (human interface device) handling.

The project uses code and components created by RealRobots and NickZero. This tutorial assumes that you'll be programming Pico from a Debian-based Linux system such as Raspberry Pi OS. However, all the hardware and microcontroller programming elements are entirely platform-agnostic.

01 Prepare your hardware

We built this project on breadboard. If you want to do the same, it's a good idea to use a serial port with a screw terminal block – both open (**magpi.cc/dbusblock**) and closed (**magpi.cc/db9block**) designs are available. Attach Raspberry Pi Pico - with headers - to a small breadboard. We like Monk Makes' Breadboard for Pico (**magpi.cc/breadboardpico**) because of its built-in GPIO cheat sheet, but any breadboard works perfectly well. See step 12 if you want to

make the project a little more permanent with a case and soldered connections to a standard male D-sub port (**magpi.cc/connectorsubd**).

02 Connect the DB-9 port

You'll need six male-male DuPont jumper cables to connect the port. We don't have any specific feature requirements in our breadboard build, so you can choose whichever of Raspberry Pi Pico's GPIO pins you fancy. If you're encasing the build in any way, pick pins that won't get in your way.

As you'll see in our wiring diagram, we've connected serial port pin 1 (up) to GPIO2, serial pin 2 (down) to GPIO 3, serial pin 3 (left) to GPIO 4, serial 4 (right) to GPIO 5, serial port pin 6 (fire) to GPIO 6, and serial pin 8 (ground) to the ground connection on Pico's third physical pin. Finally, plug your joystick's connector into the port.

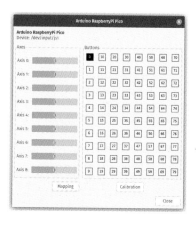

◀ Our favourite joystick tester is jstest-gtk. It is available from most Linux distros' repositories

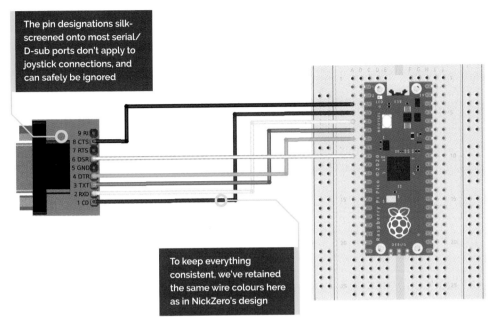

The pin designations silk-screened onto most serial/D-sub ports don't apply to joystick connections, and can safely be ignored

To keep everything consistent, we've retained the same wire colours here as in NickZero's design

9 RI
8 CTS
7 RTS
6 DSR
5 GND
4 DTR
3 TXT
2 RXD
1 CD

03 Install the Arduino IDE

Download the latest Arduino IDE from **magpi.cc/arduinoide** or, if you're running Raspberry Pi OS:

```
sudo apt install arduino
```

If you're running the IDE on Linux, including Raspberry Pi OS, you'll need to download and run as root the post-install script from **magpi.cc/postinstallsh** to create a UDEV rule allowing the IDE to write to Pico, thus:

```
sudo sh post_install.sh
```

Next, add your username to the dialout group for serial port access:

```
sudo adduser YourUsername dialout
```

04 Add the PicoGamepad library

Download RealRobots' Raspberry Pi Pico Gamepad/Joystick Library as a ZIP file from **magpi.cc/picogamepad**. Open the Arduino IDE and, from the top menu bar, go to Sketch > Include Library > Add .ZIP Library, and browse to **PicoGamepad-master.zip**. To make sure that it's been added correctly, go to Files > Examples and look for PicoGamePad under Examples from Custom Libraries.

05 Customise your DB9 converter code

Download NickZero's DB9 converter code as a text file from **magpi.cc/picoconverter** or our **DB9-joystick.ino** from our repository at

Add the GPIO numbers for your buttons

magpi.cc/usbtodb9, and open it in Arduino IDE. Where PIN_BTN_0 through 4 are defined at the beginning of a script, add the GPIO numbers for your fire, up, down, left, and right buttons. 0 is fire and the directional controls start at 1.

▼ We built our adapter on breadboard using a D-sub port with convenient screw terminals

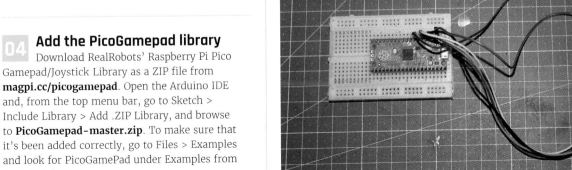

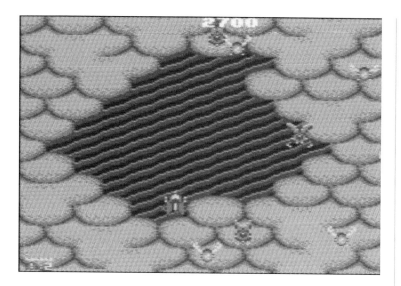

▲ Our retro joystick is ideal for playing modern C64 games like Sarah Jane Avory's Zeta Wing (**magpi.cc/zetawing**)

By default, up, down, left, and right are configured as joystick directions defined on X and Y axes. However, you can reconfigure them to fire buttons instead or as well.

06 Get ready to write

Press and hold the BOOTSEL button on Pico as you connect it to your computer. You'll have to do this every time you wish to write a sketch from the Arduino to the microcontroller. In the Arduino IDE, from the top bar, open Tools > Boards Manager and search for 'pico'. Select Arduino Mbed OS RP2040 Boards by Arduino, select the most recent version (3.3.0 at time of writing) from the pull-down, and click Install. Close Boards Manager by selecting it from from the menu again. Make sure Tools > Boards > Arduino Mbed OS

◀ The joystick is immediately recognised by the Vice emulator: all you have to do is assign it. You'll want to configure it as Joystick 2 for most games, but there's a keyboard shortcut to swap them

RP2040 Boards > Raspberry Pi Pico is selected and that Tools > Port is /dev/ttyS0.

07 Upload now

Click the upload button, a right-facing arrow, in the quick bar at the top of the IDE. If all goes well, the Output pane at the bottom of the window should show that the program has been compiled and loaded into flash.

If you want extra detail in your Output pane, open File > Preferences and select Show verbose output for compile and upload time.

If you're running Linux and getting a mysterious 'error occurred while uploading the sketch', make sure that you've run the **post_install.sh** script from step 3.

08 Test your joystick

Time to install a joystick tester. On Raspberry Pi OS and other Debian-based Linux distros, open a command-line terminal and type:
`sudo apt install jstest-gtk`.

Run jstest-gtk and you should see an Arduino RaspberryPi Pico associated with **/dev/input/js0**

⁊ Test your stick and button and see what lights up ⁊

listed. Double-click this to open the button testing screen. Now test your stick and button and see what lights up. Alternatively, visit **gamepad-tester.com** for a platform-agnostic HTML5 joystick tester.

09 Modern mappings for modern games

Now that your joystick's working, you can absolutely play Minecraft with this setup – for a given value of 'play'. We installed AntiMicroX as a Flatpak on our Linux gaming system and mapped our joystick axes to **WASD** and fire to our left mouse button. So you can move forwards, backwards, strafe, and break blocks. Turning? Who needs it? But modern games aren't really what we're going for here.

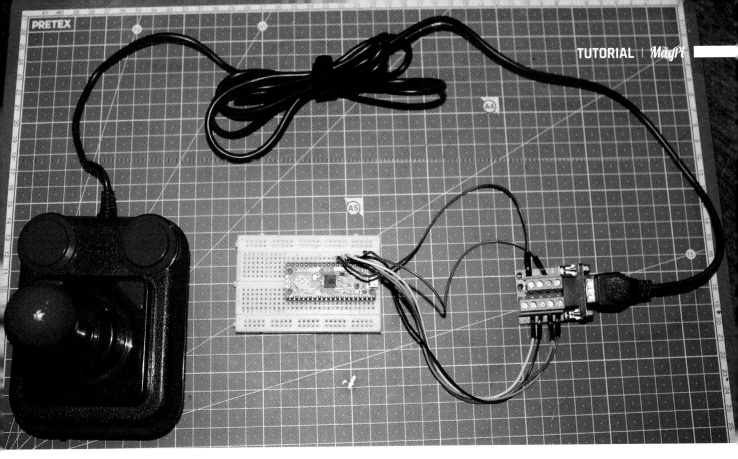

10 Retro emulation time

Install an emulator. C64 Forever and its version of Vice (**c64forever.com**) are free on Windows and you can use its licensed system ROMs with Vice on other operating systems. The Flatpak version (**magpi.cc/flathubvice**) is an easy option for Linux. You can find our guide to running Vice on Raspberry Pi in *The MagPi* 102 (**magpi.cc/102**) and find new C64 games to buy and download for free at **magpi.cc/itchc64**.

Open Vice's C64 emulator, go to Preferences > Settings > Input devices > Joystick and select the Arduino Raspberry Pi Pico. Remember, the C64's primary joystick was Joystick 2 – press **ALT**+**J** to swap at any time. Go to File > Smart attach, select your game's file, then its PRG in the pane on the right. Wait for it to load, then type RUN.

11 Extra buttons

If you're using a Competition Pro joystick like ours, a modern clone, or most standard joysticks from the C64 through the Amiga era, all the fire buttons are linked to pin 6, but you can physically mod your joystick to give you two separate fire buttons. Some two-button configurations may also require the addition of resistors to pins 9 (fire 2) and 7 (+5 V). You'll have to modify the converter script accordingly to define and invoke fire 2, for example by creating PIN_BTN_5 and supplementing the 'Button on pin' and 'Fire Button' sections of the code accordingly. You can also find contemporary two-button joysticks, but they're relatively unusual.

12 Tidying up

NickZero's make includes STL files for a case with built-in switch for a fully soldered and encased DB9-to-USB peripheral. However, you'll probably need to iterate on the Thingiverse files provided at **magpi.cc/picoconverterfiles** to make it fit the DB9 port you have. Find our iterations at **magpi.cc/usbtodb9**.

In our first print from the original STL files, the hole for the D-sub port was too small for to fit the male 9-pin serial port we had, while the switch designed to press Pico's on-board button was too wide for the hole it's supposed to slot through. The overall design is great, though: the case fits together nicely and looks professional, so is well worth tweaking if you have a 3D printer and the patience to experiment. 〽

▲ Our Competition Pro might not feel as smooth as a modern pad with analogue sticks, but it's absolutely the right tool for our retro gaming needs

▲ AntiMicroX is ideal for mapping your retro joystick to whatever preposterous purpose you may put it to, from controlling your mouse cursor to (sort of) playing Minecraft

Top Tip 👍

AntiPIcrox

If you're using your joystick on a Raspberry Pi computer, AntiMicroX is ready to install from the repository with `sudo apt install antimicrox`.

Build an arcade machine:
Get the parts

If you've ever wanted to build your own arcade machine, here's your guide.
This month: the parts you'll need, how to choose them, and where to buy them

MAKER

K.G. Orphanides

K.G. is a writer, maker of odd games, and software preservation enthusiast. Their family fully supports the idea of an arcade machine in the living room.

@KGOrphanides

Over the following pages, we'll go through the process of sourcing, building, connecting, and installing a Raspberry Pi-based arcade cabinet.

While you can restore and convert a former JAMMA cabinet for use with Raspberry Pi, or build a cab entirely from scratch, we'll be taking the flat-pack route. This lets you build the cabinet of your dreams relatively easily, somewhat cheaply, and without recourse to full-on home woodworking.

This tutorial series will use an LCD screen due to the inconvenience of sourcing and potential issues with installing a CRT model, which carries the risk of a dangerous electric shock if not correctly discharged.

01 Choose your cabinet style

If you're after a classic upright one- or two-player cabinet, then you'll want either an all-in-one model or a 'bartop' cabinet with a pedestal or stand. Bartop cabinets can also be bought without the optional stand and placed on a table. Flat 'cocktail' or 'coffee table' style cabinets are available in models for between one and four seated players and often use a vertically oriented screen, which can be split by software into two horizontal views for multiplayer games.

Other models include seated upright cabinets (often designed to take very large screens), angular tabletop models, and mini-bartops with 10-inch displays for those short on space.

02 Big screen glamour?

The size of your screen dictates the size of your cabinet, and vice versa. Before you start shopping, work out where you want the cabinet to live, and take height, width, and depth measurements. If you're working with a 19-inch monitor, you'll likely get a bartop cab that's a little under 50 cm wide. This is the most practical choice if available space is limited. A 22-inch screen translates to a cabinet of a little under 60 cm, and a 24- or 25-inch screen means a cabinet width of a bit under 65 cm. You're generally fine fitting a smaller screen in a larger cabinet, but the end result won't look quite so polished.

Check the internal measurements of the cabinet against those of the monitor, including its bezel.

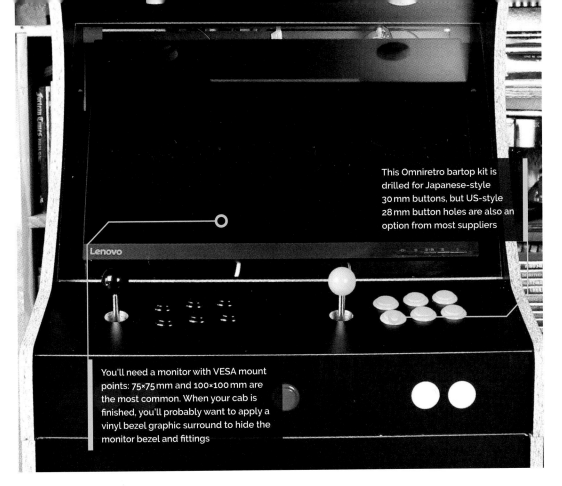

This Omniretro bartop kit is drilled for Japanese-style 30 mm buttons, but US-style 28 mm button holes are also an option from most suppliers

You'll need a monitor with VESA mount points: **75×75 mm and 100×100 mm** are the most common. When your cab is finished, you'll probably want to apply a vinyl bezel graphic surround to hide the monitor bezel and fittings

Top Tip 👍

Button positioning

We're going with a six-button Japanese-style layout. Check out **magpi.cc/ joysticklayout** to see some alternatives.

03 A good fit

Depending on the era of games you want to play, a large 1920×1080 widescreen display may not be the most authentic choice, but it is the most flexible, and modern emulators handle HD displays well.

Most cabinets have a VESA mount, usually in the form of a monitor support bar drilled for 75×75 and 100×100 mount points. Make sure your monitor has mounting points that match.

Finally, ensure that your monitor will work with Raspberry Pi: anything with a standard HDMI input should be fine, but older DVI and VGA displays require inconvenient adapter arrangements.

04 Materials

Self-assembly cabinets are usually made in MDF, but laminate, melamine, and veneer finishes are also widely available.

MDF swells badly if exposed to water, so if you're going to have drinks anywhere near your cabinet, a water-resistant finish is strongly recommended. If you buy an untreated MDF kit, apply and sand down between multiple coats of an MDF-specific solvent-based primer, then paint it to your heart's content, ideally with oil-based paint.

18 mm MDF is common, but you'll find cabinets in anything down to 10 mm for budget models. 18 mm or thicker construction materials may require a longer shaft or extender for your joystick. If in doubt, talk to the kit's supplier.

05 Finish and decoration

Regardless of the materials used, you'll probably want some plastic edging strip. This plastic trim helps to protect the edges of your cabinet, makes it easier to clean, and looks a lot more professional than exposed MDF edges.

Two types are popular. T-Molding is more secure but requires a slot to be cut for it to clip into – some DIY kits have ready-cut slots for this purpose, but budget models frequently do not.

U-Molding just clips over the edge. Cabinet makers will usually tell you how much moulding

▼ You can get kits containing all the joysticks, buttons, and connectors you'll need: just make sure your button and cabinet hole sizes match

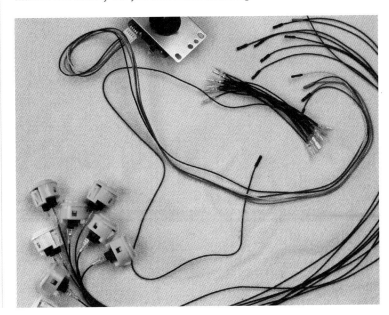

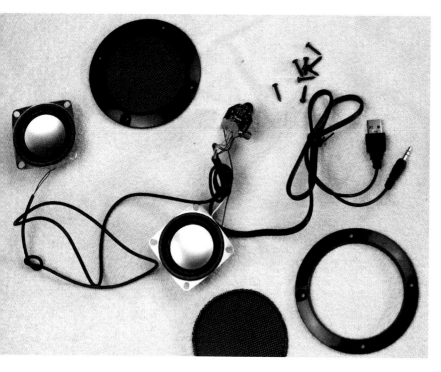

A variety of compact bus- and mains-powered amp and speaker kits are available: this one takes power from the USB port and audio from the 3.5mm port

Sample shopping list

Here is an illustrative price list. The prices include VAT but not shipping or additional costs.

Item	Price
24-inch LCD monitor	£125.00
Bartop cabinet	£170.00
Bartop stand	£100.00
10 m T-Molding	£25.00
Acrylic control panel guard	£25.00
Two-player USB joystick + button kit	£70.00
Amp, speaker & cover kit	£25.00
Amp power supply	£12.00
Printed marquee	£6.00
LED strip lighting	£15.00
Molex power adapter for LEDs	£15.00
5-way plug bar	£15.00
TOTAL	**£603.00**

their kit will need and can usually supply the required quantity and type of edging.

Many arcade cabinet suppliers also sell a range of decorative and protective graphical vinyl sticker wraps. These should be applied with care to an appropriately finished surface (check with the sticker manufacturer for any finish requirements).

06 A giant screen protector

To protect your screen and create a flush finish, you can – and should – opt for an acrylic (polymethyl methacrylate, also known as plexiglass) screen protector. Again, this is something most self-assembly kits are designed to take and the majority of retailers will happily sell you one as either a standard part of the kit or an optional extra. Make sure you do opt in, as cutting your own plexiglass to precise dimensions can be a pain. Toughened glass and UV-resistant polycarbonate can also be used. You may need to add some standoffs to stop front monitor buttons being pressed by the screen protector.

07 The marquee club

Also included in kits as a matter of routine is a strip of acrylic for your cabinet's top marquee. You'll probably want to get a backlight-ready vinyl marquee (available from print shops, arcade suppliers, and on Etsy) to stick to this, but you could also decorate your own.

While you're at it, you may wish to get acrylic or metal panels to surround your buttons and joystick. These can be decorated, and protect your cabinet's surface, as well as providing a smoother feel. Button layouts tend to be standard, but these should ideally be bought from the same supplier as your kit for the best fit.

08 Raid the button tin

We'll be building a cabinet with an eight-way joystick and six 30 mm buttons, plus Start and Select buttons, for each player. A variety of alternative sizes and brands are available, with Sanwa perhaps being the most recognisable. You can order a cabinet with holes for extra side buttons if you're into digital pinball.

An easy cross-platform connection solution is a USB arcade encoder. Models by Zero Delay and Xin-Mo are popular, but the I-PAC 2 keyboard encoder has slightly lower latency.

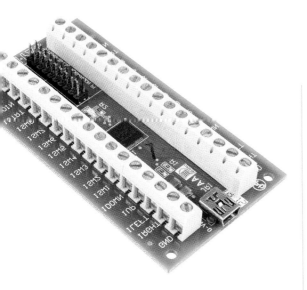

If you want to use USB, the Ultimarc I-PAC 2 encoder is a popular choice that'll work with most computers. Check out **magpi.cc/ultimarcgit** for advanced configuration

09 Pick a driver

You can connect controls to Raspberry Pi's GPIO, using either the Adafruit Retrogame (**magpi.cc/adaretrogame**) or mk_arcade_joystick_rpi (**magpi.cc/mkjoystick**) drivers – we'll be using the latter.

Arcade joysticks generally use a five-pin JST connector, while non-illuminated buttons each have a pair of quick-connect spade connector fittings, one of which must go to ground. Spade to DuPont GPIO cables are uncommon, but can be bought either individually or as part of a kit from specialist retailers such as SmallCab. Illuminated button kits are available with an extra external PSU.

LED strip lighting is a popular choice for marquee panels

10 The sound of success

It's a good idea to order your cabinet with a couple of pre-drilled speaker holes and covers to go over them. The most common option for audio is an externally powered stereo amp, connected to Raspberry Pi's 3.5mm port, and 10cm/4-inch speakers, but USB-powered kits are also available. If you have one lying around, you could also consider mounting a compact USB sound bar behind your speaker grilles.

11 More power, Igor!

A major advantage of this kind of arcade machine build is that there are no internal power supplies to bother with. There's enough space to mount a plug bar inside most cabinets, and you can use this to power the monitor, Raspberry Pi, and any extra transformers required for lights or speakers.

Where to buy

There are a number of UK and EU retailers specialising in self-assembly arcade cabinets and components. While it's easiest to get everything in one place, you have to mix and match for specialist components such as GPIO-compatible wiring looms.

- **Arcade World UK** – arcadeworlduk.com – supplies a wide range of kits and components; discount codes available for most non-furniture items
- **Bitcade** – magpi.cc/bitcadekits – UK arcade machine maker that also supplies kits
- **Omnireto** – omniretro.com – Spanish firm with a notable budget range
- **Rockstar Print** – rockstarprint.co.uk – custom marquee and wrap printer
- **SmallCab** – smallcab.net – French supplier of arcade kits and hardware including GPIO-friendly wiring

LED strip lighting is a popular choice for marquee panels, but you'll need to buy a Molex power adapter to go with it, or repurpose a PC power supply. You can run a plug lead out of the back or optionally install an external power socket and switch, if you're comfortable with simple electrical wiring.

12 Room to build

Before you start ordering, consider not only the space you have to house your cabinet, but also how much room you have to build in. Don't get an untreated MDF cabinet unless you have a large, ventilated (and paint-resistant!) space where you can apply primer to each part, as well as appropriate eye and breathing protection. ◼

Warning!
Paint and dust

When sanding, sawing, or painting, be sure to use appropriate eye and breathing protection in a well-ventilated space.

magpi.cc/diysafety

You'll want to source durable joysticks and buttons for your arcade machine

Build an arcade machine:
Assemble your cabinet

Once your arcade cabinet kit arrives, it's time to put everything together

In this tutorial, we will assemble an arcade cabinet, fit controls, and mount a monitor. You should follow the video or written assembly instructions for the model you buy, but we'll go through the process so you know what to expect and how to handle the awkward bits.

Kits don't necessarily come with the screws and bolts you'll need to attach parts such as speakers, speaker grilles, and monitors, so check that you have all the hardware you'll need before you start.

Our cabinet is an Omniretro Bartop Arcade King with a stand (**magpi.cc/kingbartop**), made of 16 mm black melamine laminate, and we are using a 24-inch monitor.

01 Lay out your parts

MDF and melamine laminate are light, cheap, and sturdy when assembled, but can be susceptible to damage if dropped or pivoted hard on an edge or corner.

Make some space and put down towels to protect the cabinet parts and your floor from one another. If your unit consists of a separate bartop and stand, build them one at a time. Read or watch the manufacturer's instructions and make sure that you have all parts, fixings, and tools to hand before you start.

02 Preparation

Assembly varies from brand to brand. If access to the assembled cabinet is restrictive, you may have to fit your buttons and joystick to it before you put it together.

Similarly, attach speakers to the inside of the marquee bottom and speaker grilles to the outside before you assemble the cabinet. If you're working with laminate, mark up the screw positions through their holes with a paint pen and use a 3mm bit to drill pilot holes. If you've already decided on your marquee, control panel and bezel graphics, your life will be easier if you apply these to their acrylic sheets before assembly (we'll be looking at this in detail in a later tutorial).

> Put down towels to protect the cabinet parts and your floor from one another

03 Assembly

If you're comfortable with self-assembly furniture, an arcade cabinet shouldn't present too much trouble, but a second person can be helpful for fitting and moving awkward parts.

Ours has a control panel with a hinged access door beneath it, so we attached this hinge first

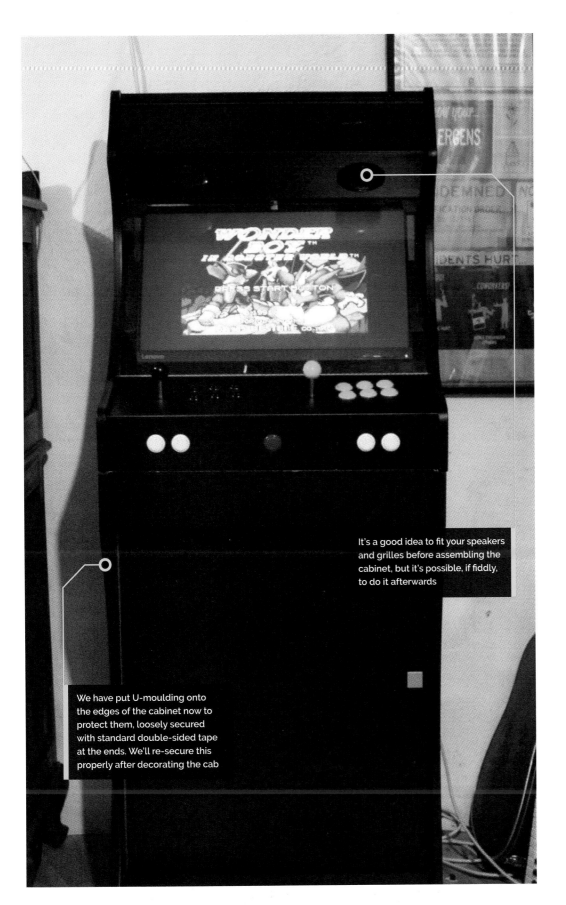

It's a good idea to fit your speakers and grilles before assembling the cabinet, but it's possible, if fiddly, to do it afterwards

We have put U-moulding onto the edges of the cabinet now to protect them, loosely secured with standard double-sided tape at the ends. We'll re-secure this properly after decorating the cab

Top Tip 👍

Snap-out

Snap-in buttons can be hard to remove without damage. The ButterCade Snap-out tool for push-buttons (**magpi.cc/ snapout**) is a plastic device to help with this.

You'll Need

- Screwdrivers, spanners, Allen keys, crimping tool

- Cordless drill

- Drill bit set. Screwdriver bits, drill bits, countersinks, tank cutters

- Additional bolts, screws, female spade connectors (to mount components)

- Dremel (recommended) and 3mm drill

- Paint pen (silver if you have black laminate, black for MDF)

- Old towels or sheets to protect parts

- Foam cleanser and microfibre cloths (to clean your cabinet and acrylics)

The underside of a Sanwa JLF-TP-8YT joystick. Note the e-clip securing the central shaft

then set the panel aside. We then attached the hinge for the bartop's rear access door and base, lined this part up with the cabinet's hood-like top, and bolted all of these parts to one side panel laid on top of them.

Lining bolts up with pre-drilled holes for this kind of build can be fiddly. If you have trouble, screw the bolts through the side panel until they're protruding, and use them to help find the correct positions.

The control panel
04
With one side now in place, slide in the control panel and bolt it to the same side as the other parts. Next, attach the marquee bottom that houses the speakers, which should already be mounted at this point.

With this model, we then close the latch on the rear access door and carefully flip the entire cabinet over onto the now-secure side panel. This is the best time to slide the marquee and screen acrylic panels into place. If you've not already applied graphics to them, leave their protective film on – it's easy to peel off later.

We now position the second side panel. We recommend again screwing in the bolts until they just protrude from the opposite side to help you lower the panel securely and accurately onto its pre-drilled holes.

Feel the power
05
Drill a hole at the back of your bartop and run the power bar's cable out through it to connect directly to a plug socket.

Some suppliers will wire a socket and bar for you, but note that international plug standards differ. Use a plug bar that can be surface-mounted inside of the cabinet.

While you're back there, cut a hole to accommodate a booted Ethernet cable or, more tidily, a screw-down Ethernet extension port. This will make Steam Link game streaming easier.

Extending your shaft
06
If your cabinet is over 16 mm thick, you'll want a longer than standard joystick shaft. Shafts are easy to swap, but watch out for parts dropping out.

Like most sticks, our Sanwa JLF-TP-8YT's shaft is held in place at the bottom by an e-clip. Hold the unit upside down, press on the bottom of the shaft with your thumb, and use a small flat-head screwdriver in your other hand to pull the clip off, using the slots in it. Pull the old shaft gently out from the top and push the new one in, carefully setting the pivot at the top and the spring and black plastic actuator at the bottom into place.

Use a thumbnail to depress the actuator and slide the e-clip back into place. You can also use pliers or your screwdriver to help push it on. For a demonstration, see this YouTube video on changing joystick shafts: **magpi.cc/joystickshafts**.

To fit the VESA mount, place the cabinet face-down, then put the mounted monitor face-down on the front acrylic screen. Use a tape measure to help with positioning

Cable tidy

Cable lacing is a cable management technique where a nylon cord is used to bind wires together. It can be used to create incredibly neat builds, like this Arcade Stick by Gordon Hollingworth, Raspberry Pi's Chief Product Officer.

Gordon learnt to cable-tidy this way as an apprentice for the MOD. "Tying the knot has to be done in a very specific way to avoid it looking untidy," he tells us, "basically a capital offence in the apprenticeship!" Gordon's cables have knots regularly at 1 cm, which keeps them smart. "We learnt this way because when you put a box into a plane or tank with some equipment in it, the vibration will shake apart pretty much any connection in the first hour. So this was the way it was done when electronics was more about wires connecting things than PCBs."

You can buy nylon cord and learn more from RS Components (**magpi.cc/cablelacing**).

▲ Snap-in buttons are held in place by plastic clips. Connect your DuPont GPIO cables first to make internal wiring easier

> " There's room to slide the screw slots on most joystick mounting plates "

07 Installing your joystick

Two plastic dust washers come with Sanwa joysticks. Slide one onto the shaft before you mount the stick onto the underside of your control panel.

When mounting your joystick, position it, mark up the position of the top right screw-hole on the joystick's baseplate with a paint pen, and drill a pilot hole, being careful not to go all the way through. Attach your joystick by that screw, make sure it's centred, and mark up the next hole or

holes. There's room to slide the screw slots on most joystick mounting plates, so you've got a bit of wiggle room when it comes to the final fit.

Don't worry too much about the orientation of your joystick – position it where it won't get in the way of the rest of your wiring. These are nominally designated up, down, left, and right positions; you can reassign these through wiring and in software.

Finally, slide the second dust washer onto the shaft on the other side and screw the joystick's ball on.

08 World of buttons

Snap-in buttons are ideal for thick wooden cabinets – plastic clips hold them in position inside the holes drilled for them. If you have an acrylic cover for your control panel, the buttons will hold it in place.

It's a good idea to attach your spade connectors to DuPont GPIO jumper cables before installing them, but you'll have to connect the shared ground cable after they're in place. We wired GPIO to the right and shared ground to the left connector on each button, but it doesn't matter which goes where. Where you have longer stretches between buttons, skip a connector on the ground chain to give yourself some extra cable to play with.

You can label the end of each GPIO cable for later ease of connection to Raspberry Pi, but they're not too hard to trace in most cabinets.

Top Tip

Foot the bill

To help the cabinet stand on an uneven floor, you can fit four rubber feet to its underside.

To make it easier to line up the sides of your cabinet with their pre-drilled recieving holes, partially screw in each bolt until a couple of millimetres protrude on the far side

> **If you find that you now can't reach or fit a part, don't panic**

Before construction, lay out the parts of your bartop on some old towels to protect them

09 Bolt screen to VESA mount

Screen mounting can be fiddly. Most cabinets come with a baton-like wooden VESA mount that's designed to be screwed into place from the inside. Start by bolting your monitor to the mount. Unless you're working with a specialist cab designed for giant screens, you'll be using a 75mm or 100mm VESA mount. These usually take M4 bolts and have a depth of 10mm. So if bolts aren't included, you'll need four, at a depth of 10mm plus the depth of your mount, although you can get away with shorter if you countersink them.

10 Screw VESA mount to the cabinet

Place the bartop face-down on the ground, protecting it with a towel. Take the protective plastic off the acrylic on the inside of the cabinet. Clean the acrylic with a microfibre cloth and anti-static foam cleanser

Lay the monitor, attached to the cabinet's VESA mount, face-down on the acrylic inside your cabinet. On the interior sides of the cabinet, mark up the position of the holes in the brackets on each edge of the VESA mount. Remove the mount, drill pilot holes, then replace and screw down the display and its mount.

If your monitor has a front power button, you can use adhesive chair leg floor protectors as soft spacers to stop it from being pressed by the acrylic screen.

11 Don't panic

If you miss a stage in your build and find that you now can't reach or fit a part, don't panic. Speakers – and any other components in need of securing – can be attached internally using strong double-sided foam tape.

Most external parts can be drilled and fitted in situ. If you want to deal with decoration last, then you can sometimes pop out your acrylic panels or, better, remove one side and reattach it.

As you'll see from the photos, we have temporarily applied U-moulding to protect the edges of the cabinet. U-moulding is easy to remove and refit or replace, assuming you don't glue it down, but T-moulding is a little harder to remove cleanly.

We're now ready to connect Raspberry Pi. That will be covered in the next tutorial.

Build an arcade machine:
Command and control

We've assembled our cabinet. Now it's time to put
Raspberry Pi to work with the Recalbox arcade OS

With our arcade cabinet built, it's finally time to get emulating with Raspberry Pi. We're using the Recalbox emulation distro for this project, which has excellent GPIO arcade controller support, a slick front end, and a handy web interface to help you configure and manage it.

The RetroPie distro is a popular choice for arcade machines, and adds Steam Link support, but requires manual installation and pull-up switch reconfiguration to get GPIO arcade controls working.

01 Wire up your controls

Last month, when we added buttons to our cabinet, we recommended attaching

the spade-to-DuPont cables that will connect to Raspberry Pi's GPIO before inserting the buttons. If you didn't, it's time to open the back of your cab, grab a torch, and get in there to fit them.

Connect a spade-to-DuPont cable to each button and connect a shared ground cable to each of the left and right button banks. Where you have longer stretches of buttons – for example between the central hot button connected to player one's rig and the player one start button – it's a good idea to skip a connector on the ground chain to give yourself some extra cable to play with.

Plug the 5-pin cable into the joystick. Looking at our Sanwa stick from below, the bottom-most pin, which connects to the black cable strand on standard-coloured 5-pin wiring harness, is ground.

02 Connect to GPIO

This is the fiddly bit. We suggest using a case for Raspberry Pi that fully exposes the GPIO pins. The GPIO wiring diagram shows which buttons, directional controls, and ground connections should be attached to each pin. While buttons and controls can be reconfigured in software, ground cannot. Our setup uses a total of 25 GPIO inputs, plus four ground connections. Input 25 is for a dedicated hotkey button.

03 Install and power up

Open Raspberry Pi Imager, connect your microSD card writer, and Choose OS > Emulation and game OS > Recalbox and the version of Recalbox

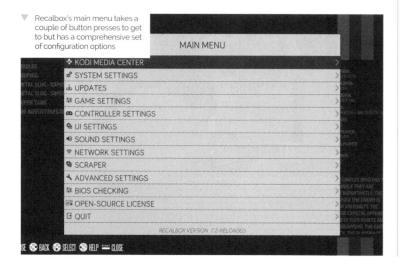

▼ Recalbox's main menu takes a couple of button presses to get to but has a comprehensive set of configuration options

MAIN MENU

KODI MEDIA CENTER
SYSTEM SETTINGS
UPDATES
GAME SETTINGS
CONTROLLER SETTINGS
UI SETTINGS
SOUND SETTINGS
NETWORK SETTINGS
SCRAPER
ADVANCED SETTINGS
BIOS CHECKING
OPEN-SOURCE LICENSE
QUIT

RECALBOX VERSION 7.2-RELOADED

BACK   SELECT   HELP   CLOSE

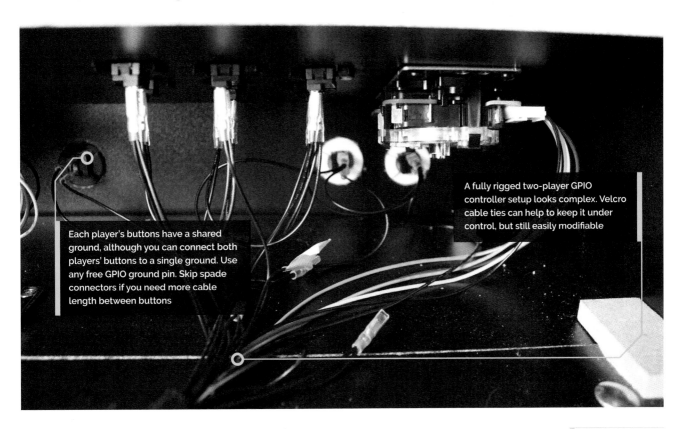

Each player's buttons have a shared ground, although you can connect both players' buttons to a single ground. Use any free GPIO ground pin. Skip spade connectors if you need more cable length between buttons

A fully rigged two-player GPIO controller setup looks complex. Velcro cable ties can help to keep it under control, but still easily modifiable

that matches your Raspberry Pi. Click Write and wait for the image file to be written to the microSD card. When Imager has finished, remove the microSD card and insert it into the Raspberry Pi in your arcade build. Connect the cabinet's monitor and speakers to Raspberry Pi. Plug in a keyboard on a long cable. Plug in Raspberry Pi's power and it will boot to Recalbox's EmulationStation interface, which you can immediately navigate using the keyboard. However, we still have to enable our GPIO arcade controls, wireless networking, and other configuration options.

04 Connecting Recalbox

Recalbox has SSH and Samba enabled by default, as well as a web interface available via your browser on **recalbox.local**. Recalbox should appear on your network as RECALBOX (File

> ❝ Recalbox has SSH and Samba enabled by default, as well as a web interface ❞

Sharing). A wired Ethernet connection will give you immediate access to these. If you don't have one, press **ENTER** to open the menu, scroll to Main Menu, and select it with **A** on the keyboard, then select Network Settings, enable WiFi, select your SSID, and then select 'Wifi Key' to enter your password. Recalbox only has a root user. The default username is **root** and the password is **recalboxroot**.

05 Configure Recalbox

You can access Recalbox's config file – **recalbox.conf** – by connecting via SSH, by browsing to the system directory in the Recalbox (File Sharing) Samba share, by pressing **F4** and then **ALT+F2** at the cabinet to exit to the console, or by going to **http://recalbox.local/** and selecting the **recalbox.conf** tab in the left-hand menu pane.

Under 'A – System options, Arcade metasystem', remove the semicolon that comments out `emulationstation.arcade=1`. This will make the arcade category the first entry in Recalbox's EmulationStation interface.

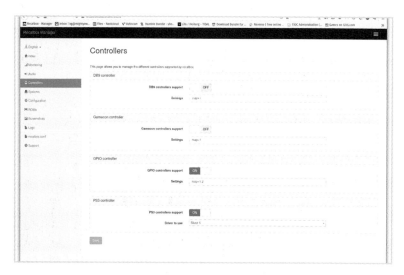

▲ A web interface at http://recalbox.local gives you control over almost every aspect of your arcade machine's setup

Top Tip

USB controls

To convert your controls to USB, use a Xinmotek board (**magpi.cc/xinmotec**) instead of connecting to GPIO.

Under D2 – GPIO controllers, set `controllers.gpio.enabled=1`. Save your changes and, at the arcade cabinet, open the menu, go to Quit > Fast restart Recalbox.

06 Optional: Take control

Recalbox will now automatically detect GPIO controllers and, if all your buttons are wired as it expects, will already have the correct button configurations. Use the bottom-left button (B) to select options and the bottom-centre button (A) to go back. Left and right navigate between systems; up and down navigate between games or options within a menu. Press Start to open the configuration menu.

If your buttons aren't connected in that order, or if you prefer an alternative layout, open the menu and go to main menu > controllers settings > configure a controller. Press down to skip an entry that you don't have buttons for. If you don't have a hotkey button for one or more players, set it to Select.

07 Sounds good

If you have no sound, open the menu, select sound settings, and check the output device. We had to switch to 'headphones – analog' output to use our cabinet's speakers, connected to the 3.5mm output on Raspberry Pi.

Recalbox plays background music all the time by default. This is charming, but a bit much for a cabinet that lives in the sitting room. Switch the Audio Mode to 'Video sound only' – to only hear the splash screens on boot – or 'No sound'.

If you prefer, you can add your own music by copying it to Recalbox's **share/music** directory.

08 Getting to know Recalbox

Recalbox comes preloaded with a number of freeware and open-source games. Because we enabled arcade mode, this category appears first. There are already four games loaded into it.

Select the category by pressing button B and scroll through them with the joystick. Gridlee, released in 1982, looks great for the era. Press B to load it.

Press Select to add credits and press Start when you're ready to play. When you've had enough, press the hotkey button and Start together to quit back to the Arcade menu.

> ❝ Press the hotkey button and Start together to quit back to the Arcade menu ❞

You can press A to go back to the top menu, and use the joystick to navigate up and down through the list. But it's easier to use the right and left directional controls to navigate through each console's full library.

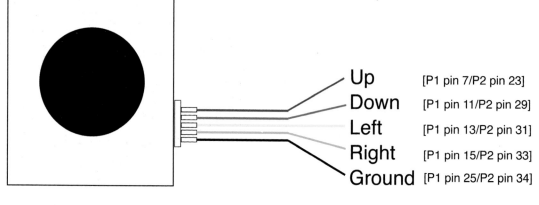

▶ Viewed from below, a standard Sanwa joystick's 5-pin connector goes to up, down, left, right, and ground. The diagram shows standard colour coding

Up [P1 pin 7/P2 pin 23]
Down [P1 pin 11/P2 pin 29]
Left [P1 pin 13/P2 pin 31]
Right [P1 pin 15/P2 pin 33]
Ground [P1 pin 25/P2 pin 34]

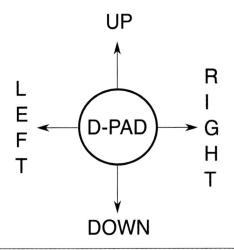

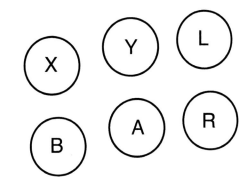

Button and joystick correspondences for player controls. The joystick maps to the D-pad. L and R correspond to L1 and R1, equivalent to the shoulder buttons of modern joypads

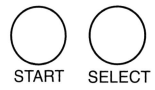

UP

LEFT   D-PAD   RIGHT

DOWN

X   Y   L

B   A   R

START   SELECT   HOTKEY

Warning!
Mains electricity
& power tools

Be careful when handling projects with mains electricity. Insulate your cables and disconnect power before touching them. Also, be careful when using power tools during this build.

magpi.cc/drillsafety
magpi.cc/
electricalsafety

09 recalbox.local

Once your arcade machine is connected to your local network, you'll be able to access it in a web browser via **http://recalbox.local**. On the main page, you'll see shortcuts to a virtual gamepad, keyboard, and touchpad, which allow you to navigate through the arcade machine's menus remotely.

To add some authenticity to older titles, go to Systems and set the Shader set to Retro, which will apply community-favourite shader and scanlines settings to each game. On the other hand, if performance is poor, disable shaders and rewinding here. Click Save at the bottom of the page to store your changes.

Below, the Configuration tab lets you set networking options, enable and disable the Kodi media player, and configure the behaviour of the EmulationStation front end and the hotkey.

10 Manage game & BIOS files

The easiest way to manage your game ROMs on Recalbox is via the web interface, where the ROMs tab lets you select the directory for your desired console, stop the EmulationStation front end, upload games, and restart EmulationStation to load them.

You can also copy games over to the **roms** directory in Recalbox's Samba share. Even if you

don't plan on emulating a specific console, don't delete the containing folders for its games, as they're required.

Recalbox also shares a **bios** directory, where you can add freeware or legally purchased computer or console BIOS files.

11 Buy and install a game

ROMs and a functional BIOS set for a number of Neo Geo maker SNK Corporation's games are available to buy as part of the Neo Geo Classics Collection (**magpi.cc/neogeoclassics**).

You'll need a Windows, macOS, or Linux PC to install or extract these. You'll find the ROM and BIOS files in the install directory; for example, **ironclad.zip** and **neogeo.zip** respectively for the fantastic scrolling shoot-'em-up Ironclad. If you don't want the whole collection, you can buy Ironclad alone at **magpi.cc/ironclad**.

Connect to Recalbox via SMB and copy the game ROMs into **roms**, and **neogeo.zip** into **bios**.

Restart EmulationStation and you should find your new games in the Arcade game list. Not all of them will work out of the box. Start any of them and press the hotkey and B buttons to open the Libretro emulation interface. Scroll down and select Options > Neo-Geo mode > Use UNIBIOS Bios. We aren't using UniBios here, but the file supplied by SNK is compatible with this setting.

Top Tip

Preconfigured USB support

If your cabinet uses a USB controller board, then RetroPie won't need any extra drivers to detect your controls.

SNK has made plenty of its arcade ROMs available to buy. Ironclad for Neo Geo-based arcade machines is a particular favourite

GPIO wiring: Connect your joysticks and buttons to Raspberry Pi's GPIO as shown. *Image by digitalLumberjack of the Recalbox project, licensed under GPL2*

and Namco – require an additional extraction and re-bundling stage. You can find tools and game lists to help you buy and use these at RED-project (**magpi.cc/redproject**) and SF30ac-extractor (**magpi.cc/sf30ac**). Linux GOG users may also require innoextract (**magpi.cc/innoextract**). Non-Neo Geo arcade games should go into the **roms/MAME** directory.

The homebrew scenes for arcade games tend to focus on physical releases, but we've had luck with Codename: Blut Engel for Neo Geo and Santaball (**magpi.cc/neogeohomebrew**) for Neo Geo CD.

For more retro and homebrew games that work well with arcade controls, including Sega's Mega Drive Classics collection, see **magpi.cc/legalgameemu** and **magpi.cc/legalroms**.

Press A twice to go back and select Resume. Your game should start.

12 Tweak your games entries

To hide the games that come with Recalbox, from EmulationStation press Start > Main menu > Games settings > Hide preinstalled games. Unfortunately, you can't pick and choose which get hidden, but you can manually download and re-add any that you'd like to keep.

You can also disable the ports category by editing **recalbox.conf** to include:

```
emulationstation.collection.ports=1
emulationstation.collection.ports.hide=1
```

If you want to add images or change the titles of the games you've added to Recalbox, the easiest approach is to use the built-in scraper. Highlight the game in the menu, press Start > Edit game > Scrape. You can also add your own ratings and keywords in this menu.

Warning!
Copyright alert!

It is illegal to download copyrighted game or BIOS ROMs in the UK without the permission of the copyright holder. Only use official purchased or freeware ROMs that are offered for download with the consent of the rights holder.

magpi.cc/legalroms

13 Get more games

The creators of the MAME emulator have been given permission to distribute some early arcade games, which you can find for download at **magpi.cc/mameroms**.

Many other emulated arcade games have been released for use on modern computers, but some – including collections by SNK, Capcom, Irem,

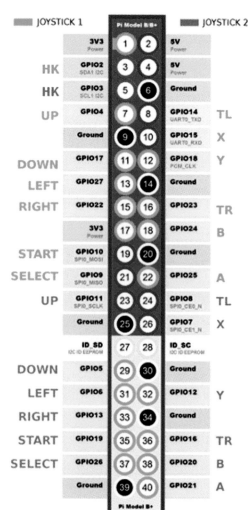

JOYSTICK 1		Pi Model B/B+		JOYSTICK 2
	3V3 Power	1 2	5V Power	
HK	GPIO2 SDA1 I2C	3 4	5V Power	
HK	GPIO3 SCL1 I2C	5 6	Ground	
UP	GPIO4	7 8	GPIO14 UART0_TXD	TL
	Ground	9 10	GPIO15 UART0_RXD	X
DOWN	GPIO17	11 12	GPIO18 PCM_CLK	Y
LEFT	GPIO27	13 14	Ground	
RIGHT	GPIO22	15 16	GPIO23	TR
	3V3 Power	17 18	GPIO24	B
START	GPIO10 SPI0_MOSI	19 20	Ground	
SELECT	GPIO9 SPI0_MISO	21 22	GPIO25	A
UP	GPIO11 SPI0_SCLK	23 24	GPIO8 SPI0_CE0_N	TL
	Ground	25 26	GPIO7 SPI0_CE1_N	X
	ID_SD I2C ID EEPROM	27 28	ID_SC I2C ID EEPROM	
DOWN	GPIO5	29 30	Ground	
LEFT	GPIO6	31 32	GPIO12	Y
RIGHT	GPIO13	33 34	Ground	
START	GPIO19	35 36	GPIO16	TR
SELECT	GPIO26	37 38	GPIO20	B
	Ground	39 40	GPIO21	A
		Pi Model B+		

LEVEL UP!

Build an arcade machine:
Decorate your cabinet

You've built an arcade cabinet, but vinyl
decals and edge moulding will bring it to life

Most arcade cabinet kit suppliers print pre-designed or custom vinyl decals to decorate your cabinet. Third-party printers can produce vinyls to your specification, but make sure that you provide accurate measurements.

Our vinyl decals, bought from Omniretro (**magpi.cc/omniretro**), arrived on a roll and had to be cut out, but some firms will die-cut vinyls for you. We'll use a wet application process, which makes it easier to remove and reposition decals for a short while after initial placement, to help you get a perfect alignment.

01 Flatten your vinyl decals

If your vinyls all came on a single roll, the first step is to cut each of them out. First separate them, if they're on a single roll, but leave generous margins. Spread them out on a table or on the floor and weigh them down – coffee table books and textbooks are good for this. Leave them for at least an hour or two: 24 hours is better.

02 Cutting out

Now they're flat, it's time to cut out your vinyls. Try to get rid of all white matter on straight edges. The easiest way is to line up a long metal ruler so that it just covers the edge of the printing, and run a scalpel down the outside of it. Curved sections for the cabinet side panels are trickier, but you don't need to worry about these as they're easy to trim down once fitted. For now, trim them freehand and leave as much white overmatter as you feel comfortable with.

03 Partial disassembly

Depending on the design of your cabinet, you may need to remove a side panel to take out the acrylic marquee and screen panels. Before doing this, use a liquid chalk pen and ruler to mark the edges of your LCD display on the acrylic, so we can accurately hide the bezel.

If you've previously fitted joysticks and buttons to your control panel, this is the time to remove them too. Apply steady pressure to the rear of snap-in style buttons to pop them out of the cabinet. People with large fingers may find a ButterCade Snap Out Tool useful for this.

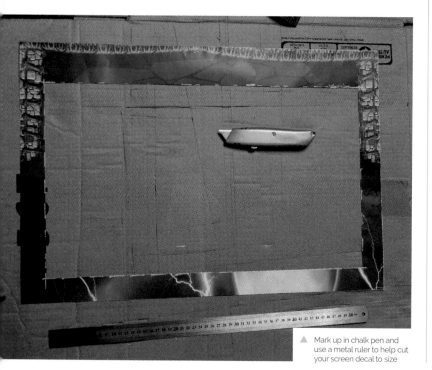

▲ Mark up in chalk pen and
use a metal ruler to help cut
your screen decal to size

04 Applying vinyl to your marquee acrylic

Two acrylic parts require individual application of vinyls: the marquee and the screen that goes in front of your monitor. The former is easy: remove the backing from the vinyl marquee decal and any protective film from the acrylic. Spray both the acrylic and the adhesive back of the vinyl with two or three squirts of application fluid. You want them to be damp all over but not awash.

Pick up the vinyl decal in both hands and, starting at one end of the acrylic, line it up with the edges and paste it down. If you're not happy with the positioning, firmly hold the vinyl and snap it back up – the application fluid will help it release easily.

Once it's positioned, use your applicator and a cloth to smooth it down, drive out any excess water, and remove any trapped air bubbles under the vinyl. Trim any excess vinyl spilling off the edge of the acrylic with a knife.

05 Measuring your screen acrylic

Cutting your screen decal to size is awkward. Before removing the screen acrylic from the cab, we marked the inner position of our monitor's bezel on the acrylic using a chalk pen. If your cabinet has a detachable VESA mount, bring the monitor with you to help line everything up.

> Grab your screen vinyl and mark up the area to cut out

Measure the distance between the edge of the acrylic and the chalk line you drew on it. Measure in multiple places to be sure of distances. Our 24-inch monitor's positioning and bezel size means that we cut 35mm in at the top and sides, and 65mm from the bottom – yours will differ.

06 Cutting your screen decal

Once you've taken the measurements, grab your screen vinyl and mark up the area to cut out. Mark on the side showing the picture, paying particular care to the corner positions. Double-check these by placing the acrylic on top to make sure both sets of marks line up.

The glowing logo is created using light-permeable vinyl on acrylic, with an LED strip mounted just behind it

A join between this bartop cabinet and its stand is rendered invisible by a large vinyl decal

Ghouls 'N Ghosts is available to buy in the Capcom Arcade Stadium collection on Steam (magpi.cc/ghoulsnghosts)

▲ Use a vinyl applicator and a cloth to stick down, remove excess moisture, and eliminate air bubbles from your decals

You'll Need

- Vinyl decals
- U-moulding/ T-moulding
- Scalpels/craft knives
- Strong scissors
- Liquid chalk marker pen
- Metal rulers, tape measures
- Vinyl application fluid
- Vinyl applicator
- Neoprene glue

Grab your metal ruler, place it along your marked line, and cut a rectangle out of the middle of the vinyl decal with a blade. If in doubt, err towards leaving too much vinyl rather than too little. To check positioning, put the acrylic over your monitor, and your vinyl over the acrylic: they should all line up.

07 Screen decal application

Now, turn the vinyl upside down, remove its backing, spray it and the acrylic with the application solution, and stick it down using an applicator and cloth. Residual chalk marks can be wiped off using a bit more of the application solution.

Allow both the marquee and screen decals to dry for a day, trim them if needed, slide them back into your cabinet, and reattach anything you removed. This will probably be the last time

you do this, so make sure the side panels are on securely and are correctly lined up and bolted to your stand, if you have one.

If you plan on back-lighting your marquee, this is a good time to put in your light. We used adhesive tape and supplied clips to mount a 50 cm USB-powered LED light on the underside of the marquee, just in front of the speakers.

08 Applying flat vinyls

If you have a full-height cabinet or a bartop and stand, you'll probably have a number of flat, front-facing areas to decorate – in our case, the front cupboard door of our stand, its base, and the front of its foot. Do these next to get your hand in.

The drill is the same for all of them: place the vinyl decal face-down on the floor, remove its backing, spray both it and the surface you're

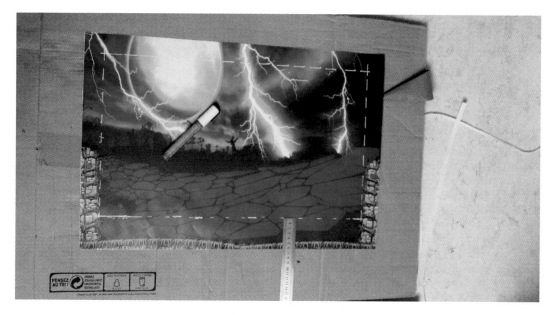

▶ We marked the inner position of our monitor's bezel on the acrylic using a chalk pen

applying it to, position your decal, and smooth it out with your applicator. Use a scalpel to trim off any overmatter. For the door, we applied the decal with the door in place – knob removed, starting at the top. We had to open the door to flatten and trim the vinyl in places.

09 Control panel decals

Most control panel decals wrap around the top and front of your panel. Buttons and joysticks should not be present during application. This is a relatively easy section to apply, but watch your position if there are decorative patterns designed to surround specific buttons or joysticks.

You may need to trim overmatter from the sides with a scalpel to get the decal to fold over the front face properly. Be careful when smoothing the vinyl on this fold, as it can be prone to both trapped air bubbles and damage from the join beneath.

10 Cabinet positioning

Side panels are the largest pieces of vinyl you'll be applying, but they're less intimidating than they seem. For a standalone bartop, one person can mount them in a vertical position with little fuss, as shown in Omniretro's video at **magpi.cc/omniretrovinyl**.

Full-height cabinets present more of a challenge due to their height and the size of the vinyl – a second person is useful here. You can apply long vinyls in an upright position, but we'd already attached rubber feet to our cabinet, so we used these to help pivot the cab down to lie on a sheet of cardboard on the floor.

11 Apply side panel vinyls

Lying flat and sprayed down as before, it's easy to line up the side-panel decal. Make sure everything's covered – with two people, it's easy to snap the decal back up if you make a mistake, then use a cloth and applicator to drive out excess moisture. Use a Stanley knife to trim the vinyl to size – its solid metal body makes it easy to follow the line of the cabinet's curves.

Go around again to remove any air bubbles and ideally leave the vinyl to dry for at least a couple of hours before pivoting the cabinet back up and lowering it to expose the opposite side. Repeat the process.

▲ You can leave some white-space overmatter on side panel decals before application, as they're easy to trim with a knife afterwards

 Side panels are the largest pieces of vinyl you'll be applying 〟

If your cabinet has separate stand and bartop parts, but uses a single sticker, there will be a slight ridge where these join. However, careful application (and a sympathetic vinyl design) makes this effectively invisible. Just be careful smoothing around it.

Top Tip 👍

Screen materials

Acrylic scratches really easily, so tinted tempered glass is an excellent alternative for your cabinet screen.

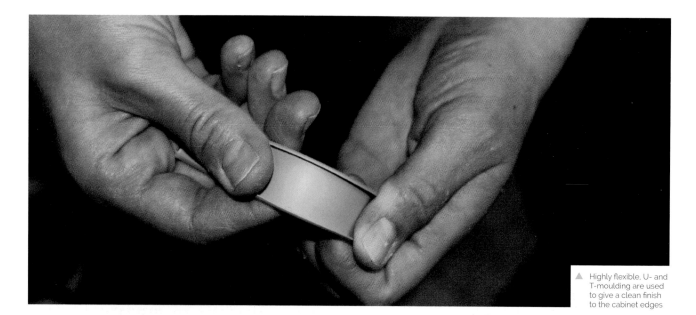

▲ Highly flexible, U- and T-moulding are used to give a clean finish to the cabinet edges

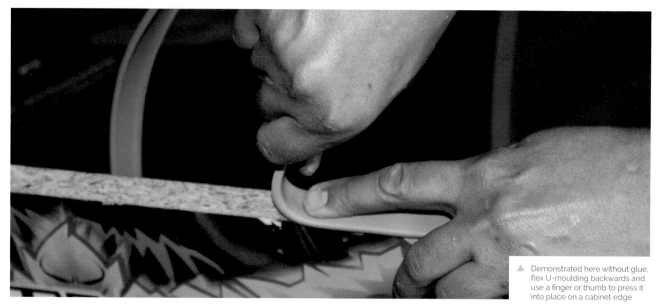

▲ Demonstrated here without glue, flex U-moulding backwards and use a finger or thumb to press it into place on a cabinet edge

Warning!
Solvents

Always use solvents in
a well-ventilated area
and keep away from
open flames.

magpi.cc/solvents

12 Moulding

We used U-moulding on our cabinet, with neoprene glue to hold it in place securely. First, measure and use scissors to cut two strips to go above and below the marquee – it's better to cut these a few millimetres too long and then trim than it is to have a gap.

Use a spatula to help apply neoprene glue along the edge you're working on, then use the tube's nozzle to apply glue to the inside of the U-moulding.

To lock U-moulding into place, bend it backwards to spread the U-shaped section, push that onto the edge you're applying it to, and then roll the moulding down along the edge, using a finger to push it into place.

When applying it to a long section, such as each side of your cabinet, start at the front underside – rubber feet help access here – apply glue to the cabinet edge and the first 50 cm of your roll of moulding, and have someone else feed it to you as you work up and around the cabinet. When you get to the bottom at the back, cut off your moulding with scissors.

T-moulding locks into a pre-cut groove along the edges of your cabinet, making it more secure, but it's still a good idea to apply glue to the flat surfaces for security. Either way, use a rubber mallet to gently tap down your moulding at the end.

You can use acetone to clean the glue off your hands and the moulding, but keep it away from the vinyl.

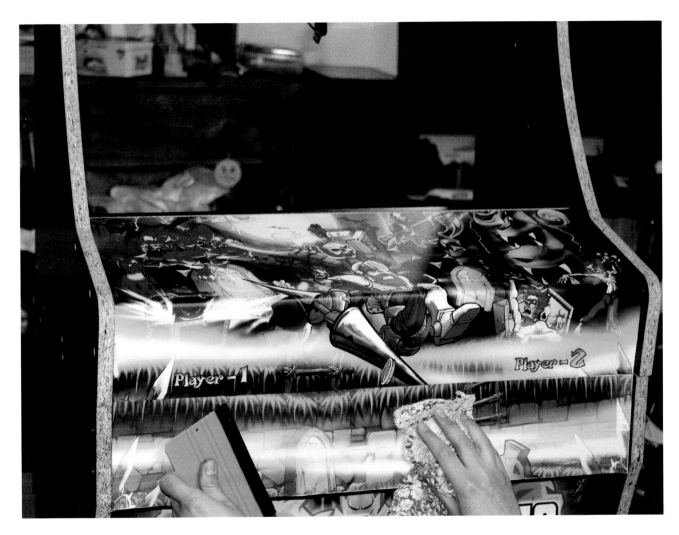

After spraying the vinyl decal, and the acrylic, with our homemade fluid, we applied it and smoothed down with an applicator and cloth

Vinyl application fluid

You can buy commercial vinyl application fluid (**magpi.cc/vinylfluid**), widely used by car customisation enthusiasts to apply decals, but we filled a spray bottle with the following homemade formula:

- 66 ml surgical spirit
- 132 ml water
- 2 drops washing up liquid

You can use warm water with a drop of washing up liquid alone, but the surgical spirit reduces drying times, which means less waiting between different stages of application and decorating.

13 Finishing moves

Use a scalpel to cut out the vinyl above the button holes: locate a hole, pierce it with the blade, slice until you find the edge of the hole, and then follow the hole round to remove all the vinyl. Do this for all your joystick and button holes.

As described in *The MagPi* #105 (**magpi.cc/105**), screw your joysticks back into place from the inside. If you're going to put protective acrylic panels over your control panel, this is the time to do it – they're held on solely by the buttons.

However, because our cabinet is for home use, we've left the vinyl bare for a more comfortable and attractive finish. If your cabinet will see lots of play, acrylic will protect it and cut down on wear and tear. Whichever you choose, connect a DuPont cable to each button and pop them into place.

Follow the instructions from issue 106 to connect your buttons and peripherals to Raspberry Pi. M

Warning!
Sharp objects

Take care when using knives and scalpels.

magpi.cc/handknives

Build an arcade machine: RetroPie and stream from Steam

Use RetroPie as your arcade operating system and add extra emulators with support for Steam Link. Stream games from a powerful PC to Rasbperry Pi

I n the previous part of this guide, we used Recalbox for our main arcade cabinet operating system, but it's not your only choice. In this final instalment of the 'Build an arcade machine' series, we'll use the RetroPie distribution, currently at version 4.8, to provide extra features such as Steam Link support, as well as taking a longer look at where to buy arcade games and how to get them onto your system. This tutorial assumes that you already have a fully assembled and wired arcade cabinet.

Operating System menu. OS customisation settings aren't available for this OS, so we'll have to manually configure those on the first boot.

While older versions of RetroPie required a manual fix to make the GPIO pullup switches work correctly, this is no longer the case in version 4.8. That means you can immediately put the microSD card into your cabinet. We'll set up everything else we need – network connectivity, SSH and some improved drivers – in the coming steps. RetroPie is based on Raspberry Pi OS Lite Buster, so we'll also tighten up security.

▼ More button assignments are available on RetroPie than you have arcade controls. You can skip the ones that don't match up

01 Install and prepare RetroPie

Fire up Raspberry Pi Imager, connect your microSD card writer, select your Raspberry Pi device and install RetroPie 4.8 for your device from the Emulation and Game OS section of the

CONFIGURING

GAMEPAD 36

HOLD ANY BUTTON TO SKIP

⊕ BUTTON A / EAST	BUTTON 7
⊕ BUTTON B / SOUTH	BUTTON 0
⊕ BUTTON X / NORTH	BUTTON 3
⊕ BUTTON Y / WEST	BUTTON 2
⬤ LEFT SHOULDER	BUTTON 4

OK

02 First boot

Make sure you have a keyboard plugged into your cabinet for this bit. Plug in Raspberry Pi's power. It should boot to the EmulationStation interface, but we can't configure a controller (other than the keyboard) until we've added support for GPIO arcade controls.

Press **F4** to exit to the command line. First, some basic housekeeping. We'll add a stronger password and enable SSH so we can do most of the rest of our configuration remotely. Type:

```
sudo raspi-config
```

Select:

```
1 System Options
S3 Password
Now enter a new passwork and return to the
top level menu. Select :
3 Interface Options
P2 SSH
Yes
```

Raspberry Pi is powerful enough to run most games, and you can stream classic and modern arcade titles from a separate PC on your network via Steam Link

Steam Link lets you create a dedicated control map. Remember to map your hotkey button to its guide button

You can now SSH into Raspberry Pi from another PC using a client such as Remmina or PuTTY.

03 Add hotkey button support

If, like ours, your arcade cabinet's GPIO controller setup has either one or two extra hotkey buttons for easy access to save, load, and exit shortcuts while playing, then the standard version of the mk_arcade_joystick_rpi driver available from RetroPie's package manager won't support them. We'll have to manually add an updated version from maintainer Recalbox's GitLab repo.

At the command line, type:

```
git clone --branch v0.1.9 https://gitlab.
com/recalbox/mk_arcade_joystick_rpi.git
 sudo mkdir /usr/src/mk_arcade_joystick_rpi-
0.1.9/
 cd mk_arcade_joystick_rpi/
 sudo cp -a * /usr/src/mk_arcade_joystick_
rpi-0.1.9/
 nano /usr/src/mk_arcade_joystick_rpi-0.1.9/
dkms.conf
```

In this file, change PACKAGE_VERSION="$MKVERSION" to PACKAGE_VERSION="0.1.9". Press **CTRL+X** to exit, then **Y** to save.

Back at the command line, enter:

```
 sudo dkms build -m mk_arcade_joystick_rpi
-v 0.1.9
 sudo dkms install -m mk_arcade_joystick_rpi
-v 0.1.9

 reboot
```

04 Optional: Load your hotkey driver

When Raspberry Pi has rebooted, SSH back in and type:

```
 sudo modprobe mk_arcade_joystick_rpi map=1,2
```

Go over to the arcade machine and press **F4** to get to the command line and test your controllers:

```
 jstest /dev/input/js0
 jstest /dev/input/js1
```

> We'll use the RetroPie distribution to provide extra features such as Steam Link support

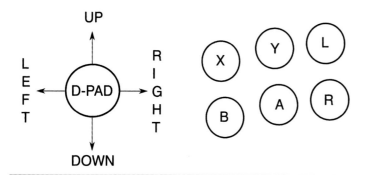

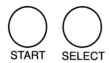

▲ Button and joystick correspondences for player controls; we recommend this configuration for use with RetroPie. L and R map to the right and left shoulder buttons

If that works, it's time to load that module on boot. At the command line:

```
sudo nano /etc/modules
```

In this file, add the following on a new line, then save and exit.

```
mk_arcade_joystick_rpi
```

Next, at the command line:

```
sudo nano /etc/modprobe.d/mk_arcade_
joystick.conf
```

In this file, add the following:

```
options mk_arcade_joystick_rpi map=1,2
```

Now save, exit and reboot.

 Configure RetroPie

There's a bit more configuration to do before RetroPie is ready to go. SSH in and type:

```
sudo ~/RetroPie-Setup/retropie_setup.sh
```

...to open the ncurses configuration menu.

If you did not manually install a hotkey version of the mk_arcade_joystick driver in the previous steps and do not need one, go to:

```
P Manage packages
driver
212 mkarcadejoystick
```

...and install it.

If you need to connect any Bluetooth keyboards or controllers, go to:

```
C Configuration / tools
198 bluetooth
```

Press **R** to register a device and follow the pairing instructions.

832 samba in the configuration menu sets up Samba shares so you can easily transfer ROMs and BIOS images over your local network

You can add extra emulators here, but we'll come to that later. For now, select the R Perform reboot option from the main menu.

Input configuration

When RetroPie reboots, it should inform you that it can detect two GPIO controllers. Press and hold any button on the left-hand button bank to configure controls for player 1. Because arcade controls don't map perfectly to a gamepad, you'll have to skip some buttons by pressing and holding any key.

Map up, down, left, and right on the arcade stick to the D-pad. Follow our button assignment diagram to map the top row to button Y, X, and L(eft shoulder), and the button below to buttons B, A, R(ight shoulder).

Map Start to player 1's left-hand front function button and Select to their right-hand front function button – this will be their 'insert coin' button. In our wiring configuration, our single hotkey button – the last we set – is associated with player 1.

> ❝ When RetroPie reboots, it should inform you that it can detect two GPIO controllers ❞

Approve your configuration, then set up player 2's controls in the same way.

Getting to know RetroPie

With your controllers configured, RetroPie's main interface will open. Press A to select menus

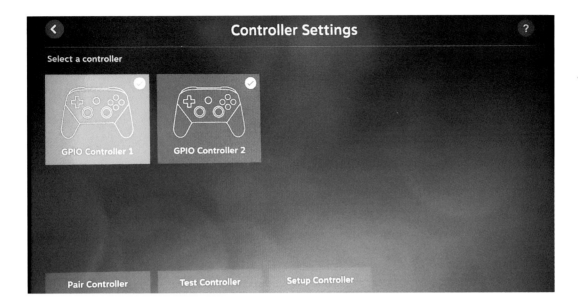

Controller Settings

Select a controller

GPIO Controller 1   GPIO Controller 2

Pair Controller   Test Controller   Setup Controller

 You'll probably want to reconfigure your controls in Steam Link to better match its Steam Controller-based expectations

and items and B to go back. Press Start to open the main menu and Select to open the options menu. Press the same button again to close each of these.

As you have yet to put any games on the system, only the RetroPie menu will be available. Here, you'll find easy access links to configuration tools, including some we used earlier. Install new emulators and drivers from the RetroPie Setup menu.

You'll probably need to disable overscan to get rid of a black border around the screen. In the ReotroPie menu, select Raspi-config > Display options > Underscan > No and then reboot to solve the problem. Note that button B is mapped to the **ENTER** key in this set of menus.

When you add any new games, ROMs or emulators, you'll have to restart EmulationStation by pressing Select, going to Quit, and then Restart EmulationStation.

08 Install more emulators

Although this is an arcade machine, you can play what you like on it. The core lr-mame2003 and lr-fbneo emulators are included, along with those for popular consoles such as the Sega Mega Drive, used in some arcade systems and for which original games are legally available.

Some emulators require system BIOS images. Sadly, very few of these have been made legally available to emulation enthusiasts. Dotemu's SNK 25th and 30th Anniversary Collection games include a UniBIOS compatible BIOS set. We recommend adding the following:

opt > 178 opentyrian – arcade-like DOS shoot-'em-up Tyrian 2.1 is now freeware.

exp > 105 lr-mame – a more up-to-date version of MAME that supports a wider range of ROMs. Install from source for bleeding edge.

exp > 158 digger – a sanctioned remaster of Windmill Software's Dig-Dug.

exp > 185 steamlink – this allows you to stream less emulation-friendly titles directly from a Steam installation on a Windows or Linux PC.

09 Configure your emulator

Once you've installed a new emulator, such as lr-mame, you'll have to configure the libretro back end to use it by default for either all games or selected titles. The easiest way to do this is to browse to the game you want to play in the EmulationStation front end.

Go to the Arcade menu, press B to start any game – it doesn't matter if it currently works or not – and then press B again when you're briefly prompted to 'press a button to configure'. Select option 1 to set the default emulator for arcade games and choose lr-mame. Option 2 allows you to select a different emulator for anything that doesn't work well with this.

10 Connect Steam Link

Linking Steam to your arcade cabinet lets you stream a wealth of modern and classic arcade games to Raspberry Pi from a more powerful

Top Tip 👍

Steam Link smoothly

Use a wired Ethernet connection for optimal Steam Link game streaming.

RetroPie is significantly more configurable than Recalbox, although its interface doesn't look quite as slick

---

Top Tip

A screw loose

If the ball on your joystick is loose, use a screwdriver in the slot on the underside of the stick or cloth-wrapped pliers to hold the shaft still while you tighten the ball.

PC, like Melty Blood, Guilty Gear, Horizon Chase Turbo, and Street Fighter V. After you've installed it and restarted EmulationStation, go to the Ports menu and select Steam Link.

It'll download updates – you will need a keyboard plugged in to approve these – and then run. Make sure Steam is running on a PC on your local network and that Enable Remote Play is ticked under Settings > Remote Play.

On the arcade machine, select the computer you want to link to. Steam Link will show a code. Enter this in your PC's Steam client when prompted. To avoid a resolution mismatch, run Steam with a monitor that matches the resolution of your arcade machine set as your primary display.

11 Configure Steam Link

You may want to reconfigure your controls, as Steam Link doesn't inherit the control layout from RetroPie's EmulationStation, and some games do better with alternative button assignments – for example, to more closely match an Xbox or Steam Controller, which swaps the position of the B and A buttons.

To set these, launch Steam Link, press up to highlight the gear icon, press A (per our button assignment diagram), go right to highlight Controller and press A. Select the controller you wish to configure, then press down and right twice, and select Setup Controller.

Hit the button you want to associate with each Steam Controller-style button as it's displayed

on screen. Use a keyboard or your second set of controls to use the skip button at the bottom to bypass extraneous buttons.

12 Why use Steam Link?

Steam Link is an invaluable tool for arcade emulation enthusiasts, not only because you can play more CPU-intensive games, but also because it's the best way of ensuring copyright compliance for a number of re-released arcade games.

We've been playing Ghouls 'N Ghosts from the Capcom Arcade Stadium on our cabinet via Steam Link. Unlike some SNK and Sega re-releases, Capcom doesn't supply emulator-ready ROM files and the EULA for that compilation doesn't allow you to extract its PAK files.

Neil Brown of decoded.legal opines (**magpi.cc/ romextractionlegal**) that "when even a legitimate Steam purchaser extracts the ROMs and runs them on their own Raspberry Pi, they infringe Capcom's copyright", making streaming these titles your best option for fully legal home arcade action. M

DOS ain't dead

You can bring new life (and games) to dead computers

The last stand-alone version of MS-DOS was released in 1994, so you might be surprised to know that a commercial DOS game came out on Steam in 2023. Hadrosaurus Software's *The Aching* (**magpi.cc/theaching**), is an otherworldly horror adventure with 16-colour graphics, an arrow key and text parser interface, and hardware support for the Tandy 1000, a PC released in 1984. You can play it on Raspberry Pi, or anything else that'll run the DOSBox emulator.

No one really uses MS-DOS any more, but the modern, open-source FreeDOS ships with every copy of DOSBox, and you've quite probably used that. Most modern DOS developers use DOSBox and its forks for testing, so they can rapidly spot bugs and iterate solutions.

The year 2023 in DOS also saw the release of Damien "Cyningstan" Walker's stylish *Barren Planet* (**magpi.cc/barrenplanet**), a turn-based, space exploitation-themed strategy game in which rival mining corporations battle for control of resources, with some of the best four-colour CGA graphics we've ever seen. Cyningstan has also released a range of tools and libraries (**magpi.cc/cyningstan**) to

support DOS games development in C, as well as open-sourcing his older games.

Juan J. "Reidrac" Martinez, developed *Gold Mine Run!* in C and cross-compiled from Linux to DOS, using DJGPP to target 32-bit (i386) DOS. He also open-sourced (**magpi.cc/goldminerungit**) the game's code to help other developers.

But you don't have to use C. Tiny DOS city-builder *Demografx*

> ## ❝ A commercial DOS game came out on Steam in 2023 ❞

(**magpi.cc/demografx**) was developed in Microsoft QuickBasic 4.5, an IDE released in 1990, which you can run on Raspberry Pi in DOSBox if you can find a copy. Microsoft's more common QBASIC and GW-BASIC languages are no longer available, but PC-BASIC (**magpi.cc/pcbasic**) is a fully-compatible GW-BASIC interpreter you can install on Raspberry Pi, and there's even a GW-BASIC extension for Visual Studio Code if you want an IDE.

There's an entire community of developers making wildly

distinct games based on ZZT (**museumofzzt.com**), a 1991 game creation system by Tim Sweeny, now CEO of Epic Games. ZZT spawned a vast living ecosystem of DOS games like *WiL's Galactic Foodtruck Simulator* (**stale-meme.itch.io/gfs**), development tools like KevEdit (**magpi.cc/kevedit**), and modding tools such as Weave (**meangirls.itch.io/weave-3**).

There are multiple DOS game jams to encourage would-be developers. In 2023, we saw the DOS COM jam (**magpi.cc/doscomjam**), the DOS Games June Jam (**magpi.cc/2023jam**), and the DOS Games End of Year Jam (**magpi.cc/dosgoty**).

The DOS renaissance still has a way to go before it catches up to the C64, ZX Spectrum, and Game Boy development scenes, but the sheer range of tools available makes it a very approachable space to experiment in. If you want some inspiration, check out this DOS games we've created at **magpi.cc/itchdos**. ▥

AUTHOR

K.G. Orphanides

K.G. fell in love with games writing and the command line in DOS and decamped to Linux in protest over Windows Me. They regret nothing.

twoot.space/@owlbear